W9-AFK-404

South-East Asian Social Science Monographs

Sovereignty and Rebellion

Sovereignty and Rebellion

The White Hmong of Northern Thailand

Nicholas Tapp

SINGAPORE
OXFORD UNIVERSITY PRESS
OXFORD   NEW YORK
1989

4/8/90

15

Oxford University Press

Oxford New York Toronto
Delhi Bombay Calcutta Madras Karachi
Petaling Jaya Singapore Hong Kong Tokyo
Nairobi Dar es Salaam Cape Town
Melbourne Auckland
and associated companies in
Berlin Ibadan

Oxford is a trade mark of Oxford University Press

© Oxford University Press Pte. Ltd. 1989

Published in the United States by
Oxford University Press, Inc., New York

All rights reserved. No part of this publication may be reproduced,
stored in a retrieval system, or transmitted, in any form or by any means,
electronic, mechanical, photocopying, recording or otherwise,
without the prior permission of Oxford University Press

ISBN 0 19 588912 6

British Library Cataloguing in Publication Data
Tapp, Nicholas
Sovereignty and rebellion: the White Hmong of Northern
Thailand.—(South-East Asian Social Science Monographs)
1. Thailand. Northern Thailand
I. Title
959. 3' 0049591
ISBN 0-19-588912-6

Library of Congress Cataloging-in-Publication Data
Tapp, Nicholas.
Sovereignty and rebellion: the White Hmong of Northern Thailand /
Nicholas Tapp.
p. cm.—(South-East Asian social science monographs)
Bibliography p.
Includes index.
ISBN 0-19-588912-6:
1. Hmong (Asian people)—Social conditions. 2. Hmong (Asian
people—Thailand. 3. Refugees—Thailand—Nomya. 4. Nomya
(Thailand)—Social conditions. 5. Hmong (Asian people)—Religion.
I. Title. II. Series
DS570.M5T37 1989
305.8'950593—dc20
89-9455
CIP

Printed in Malaysia by Peter Chong Printers Sdn. Bhd.
Published by Oxford University Press Pte. Ltd.,
Unit 221, Ubi Avenue 4, Singapore 1440

This work is dedicated to Khinthitsa,
for her constant support and encouragement

All this goes on inside me, in the vast cloisters of my memory. In it are the sky, the earth, and the sea, ready at my summons, together with everything that I have ever perceived in them by my senses, except the things which I have forgotten.

(St. Augustine, *The Confessions*, Book X)

We are committed to the conclusion that all history is necessarily written from the standpoint of the present, and is, in an inescapable sense, the history not only of the present but of that which is contemporaneously judged to be important in the present.

(J. Dewey, *Logic: The Theory of Inquiry*)

Acknowledgements

FIELDWORK was conducted from April 1981 to October 1982 with assistance from the Social Science Research Council and the Central Research Fund of the University of London for which I am grateful. I would like to express particular thanks to Dr A. Turton, my supervisor, who constantly encouraged me to follow my own interests on the Hmong, and to Dr J. Lemoine of CeDRASEMI, who has given me much valuable and needed advice and criticism on the Hmong. I am also most grateful to Dr Yangdao, previously of the University of Minnesota, whom I met before going into the field, for his unfailing kindness and sympathy. I have benefited from the comments and criticism of Professor G. Downer, the late Professor Sir E. Leach, and Dr S. Feuchtwang, and am also grateful to the British Academy through the Institute of Southeast Asian Studies, Singapore, who funded my second visit to the United States. Thanks must also go to the Thai authorities who permitted me to work in sensitive areas of their society, to Khun Chupinit Kesmanee of the Tribal Research Centre, and the personnel at the Hmong Centre in Chiangmai who aided me in the field, and above all to the Hmong villagers of Thailand, who proved unfailingly hospitable, courteous, and sympathetic to my attempts to gain some knowledge of their culture and ways of being. In particular, I should like to thank Vaj Vam Xauv, Thoj Xeeb, Lis Txoov, Lis Npis, Vaj Meej, and Vaj Suav Vaj.

Photographs are by the author and Khinthitsa.

Note on Orthography

I have throughout used the Barney–Smalley system of Romanized Phonetic Alphabet (RPA) for Hmong terms which was developed in collaboration with Bertrais (Bertrais 1964; E. Heimbach 1979). This is easy to read, provided one remembers that there are no final consonants in Hmong. Those shown, therefore, indicate tone values and should not be pronounced, while final nasalization is indicated simply by doubling the vowel; thus 'Hmoob' = 'Hmong' (pronounced in a high tone). One further peculiarity of the system is that 'x' is pronounced like (English) 's', and 's' like (English) 'sh'.

Contents

Appendices

Tables

Figures

Maps

Plates

Equivalents

Currency
20 baht = US$1.00 (1980)
 1 daim nyiaj (normal bar of crude silver) = c.3,000 baht (1981)
                                          c.5,000 baht (1980)

Opium
1 joi = 1.6 kg

Area
 1 daj = 2 outstretched armslengths
10 daj = 2 nga
 4 nga = 1 rai
 1 rai = 0.395 acres

Weight
1 kasorp (gunny bag) = 100 kg (of rice, potatoes, etc.)
1 seev   = 6 bip (*seev*) tang or poom
1 poom   = 20 litres

Length
½ daj = 2 tshim (cubits)
1 tshim = 2 dos (outstretched thumb to second finger)
1 noos (outstretched thumb to first finger)

PART I

I

Introduction

The Study

In this book I have not attempted a basic ethnography of the Hmong culture or economy, since adequate ethnographies already exist (Lemoine 1972a; Geddes 1976).[1] It was designed from the first to be a micro-study of certain aspects of White Hmong culture, in particular geomancy, messianism, and literacy. What I have done, therefore, through considering one central legend as paradigmatic of Hmong society as it is seen by the Hmong themselves, is to have attempted to relate this legend to all other aspects of White Hmong cultural and socio-economic organization.

While inevitably influenced by a social drama or narrative style of ethnography, which seemed particularly apposite here because I was seeking to introduce a specific, phenomenological notion of *temporality* into social analysis, I have produced what may be styled a 'textural' kind of analysis. This arises partly out of an earlier training in literary criticism, and is evidenced in the series of translated legends around which the thesis revolves.

However, I have attempted to avoid some of the pitfalls of a purely structuralist approach by attempting to relate the kind of thematic and systemic oppositions discerned in these legends to oppositions and contradictions which were present in the village in a painfully evident way, particularly in the economic and political spheres. In this way, I hoped to co-ordinate the nominative and operative planes of social discourse.

This has inevitably limited the space and time I have been able to devote, both to the more detailed consideration of Hmong history which I should have liked, and to the broader aspects of White Hmong life and belief. However, I felt that the argument of the book was sufficiently important, ethnographically and theoretically, to justify such an approach.

This means not only that I have tried to produce a work which may appeal to a wider audience of those concerned with, or involved in, the plight of the Hmong as victims of a historical process, but I have inevitably lent greater weight to what Leach (1960–1) termed the *third* level of social analysis, that is, the ideas and statements of actors about themselves and their society. Indeed, it does seem to me that, in the majority of cases, the analytical observations of the ethnographer are of no greater 'objective' value than the descriptions and analysis of his informants (Levi-Strauss 1963a) since they work from the same data. To pretend, therefore, to a greater scientific objectivity, on the basis of eighteen months of participant observation of a culture, than those of one's informants who grew up in that culture and have been familiar with it since infancy, is to be guilty of precisely the same sort of metaphysical solecism which informed, for example, the nineteenth-century author's assumption of omniscience with regard to the movements and ideas of his subjects.

It follows from this point of view, which after all is no more than (another) point of view, that just as the principle of indeterminacy had to be introduced into modern physics once the effect of the observer on the processes he observed was discovered (Koestler 1964, p. 241), so I felt that there would be no excuse for not introducing myself into the narrative at places where it seemed appropriate.[2] Although this is not the style of the traditional ethnography, it has been a matter of deliberate decision.

Moreover, given the very considerable resources which we have on the Hmong in various forms (missionary writings, translations of their songs and stories and shamanic rites, development reports, news articles, ethnographies, nineteenth-century travellers' accounts, and a recent flood of 'refugee literature'), I have sought to juxtapose these sources, where appropriate, with the words of the Hmong concerning both their present situations and their view of the past and future, and with my own comments and observations. Research cannot, after all, take place in a vacuum. Any additions may contribute to our knowledge, if only through their correction. What I hope will have emerged will have been a creative juxtaposition between other opinions, my own, and those of the Hmong as reported by me, by others, and by the Hmong themselves.

I hope that the net result will be found to be an original contribution to knowledge, both in terms of the notion of 'real history' (which I posit as a general theory but one which I believe that the Hmong, on account of their possession of an *oral* tradition and strongly creative spirit, exemplify in particular), and in terms of the original fieldwork, and the translations of Hmong comments, ideas, statements, feelings and opinions, legends and myths, which this book represents. A 'real' history I define as an *experienced* one, which may be made up in part, wholly, or not at all, out of events which truly occurred, and emerges in the context of the interaction between disagreements about historical 'facts' and conflicts between literate and oral accounts of specific 'histories'. This is the reason why I proffer this book, not as an authoritative view of Hmong culture and economy from any particular theoretical bent, but as a limited viewpoint, along with many other limited viewpoints, which may contribute to our total understanding of the Hmong, and the processes of social change and ethnicity with which this work is concerned. I have, in other words, tried to speak, as Wordsworth advised, in the 'real language of men'.[3]

Fieldwork Techniques

Although the Hmong are shifting cultivators, and their lineages are very wide, comprising parts of many scattered villages from Thailand to southern China, there can be no substitute for thoroughly immersing oneself in a single location if one wishes to learn anything about their economic, kinship, or belief systems. This I was able to do during the period of research. I would argue against those who assume that the Hmong village is somehow an unreal entity, on the grounds that the Hmong term for village, *zos*, properly refers to single-clan villages and not those common in Thailand where representatives of two or more clans are often present in the same location, for literally on the ground, which is where fieldwork begins, villages and other settlements are the way the Hmong organize their lives, and so I felt that peripatetic fieldwork would not lead to the sort of results I wished to obtain.[4]

There were in the survey site, however, several hamlets, and while covering altogether six separate villages of the White Hmong, ranging from a settlement with only four households to one with nearly fifty (Hapo), my study focused on one particular village, Nomya, which I refer to throughout as the focal village. Personal, economic, kinship, and ritual relations were maintained in complex networks across the membership of the different neighbouring settlements which illuminated relationships within this focal village. However, I benefited particularly through examining relationships between the larger settlement (Hapo) and the focal village, since these were of a less embedded nature than those between the various Nomya villages whose members frequently visited each other. Often I was able to check and verify data received in the focal village through comparison with that received in the larger settlement. Owing to an unresolved affair between a young Vaj clan male in the focal village and a Thoj clan girl in the larger settlement, courting relationships between the two were minimized, and I did not make as full a study of the larger settlement as I was able to do of the focal village. The exact manner in which I arranged my relationship with these villages arose directly out of the internal composition of the focal village itself, and this is described in Chapter 5.

The village itself, to which I was introduced by a Hmong from Phitsanalouk province, was primarily chosen for three different reasons. First, although it was clear that it was not unaffected by development, it had not suffered the extensive inroads that some villages close to Chiangmai, the capital of Northern Thailand, had experienced. Nomya remained small in population, traditional in structure, and was unvisited by tourists. Thus I felt it would be representative of a large number of Hmong villages in Thailand, and attitudes towards development measures would be easier to gauge. Secondly, I knew that the village produced opium, and I felt that the cultural issues with which I was mostly concerned would be more readily maintained in a village which was relatively wealthy. Thirdly, I knew that there had been some Christian contacts in the village, and I was particularly interested in the effects of the adoption of Christianity. Another feature of the village's representativeness of other Hmong villages in Thailand was that it impinged on the Mae Chaem watershed area, which until recently was a 'no-go' area for Thai government officers on account of the extensive communist insurgency in the region.

In the village I went to work in the fields, spent hours talking to villagers, and conducted a series of interviews with all household heads which were taped and had to be transcribed and translated with the help of a research assistant. I participated in weddings and funerals, more fully once I had attained a clan membership which allowed me to be seated, where appropriate, in the correct locations. This I did about halfway through the fieldwork period, delaying the ritual until I was sure that such an identification would be of more, rather than less, utility. I watched countless long shamanic sessions and usually conducted follow-up interviews with the shaman and the family who had consulted him, as well as witnessing many other household ceremonies and village meetings to which I was invited as a member of the village.

I had learned Thai during an earlier period in Thailand as a voluntary teacher and this experience first aroused my interest in the Hmong. Although this was useful in the very early stages of fieldwork, and allowed me to conduct interviews with Thai officials and members of other ethnic minorities outside the village, I made every effort to learn as much White Hmong as I could, and soon was communicating and being communicated with solely in Hmong. Thus all the village data on which this book is based were collected in Hmong. I was particularly fortunate to have had the benefit of a course of study in the phonology and structure of the language before leaving England.[5] Without an initial training in distinguishing the sounds of the language, I am sure I would have been unable to pick it up in the village. Communication in Hmong proved essential, both for eliciting trustworthy economic data, and for demonstrating my respect for the culture in which I was interested.

I went to great lengths to obtain recordings of songs, stories, and rituals, concerning the past and the history of the Hmong. Sometimes these had to be collected outside their normal context where recording was impossible owing to noise, activity, or ritual prohibition (such as after the birth of a child, when all shoes, bags, and their contents had to be left outside the house). Only a small portion of these materials has been included in this book, owing partly to reasons of space but more to the demands of the argument. I also conducted household census surveys at the beginning and end of my stay in the village, and constructed genealogies for all the families in the focal and larger settlements. I made systematic attempts to obtain interviews with all outsiders connected with the village, such as the Chinese shopkeepers, locally settled Burmese, Thai, and Karen, agricultural extension officers, voluntary teachers, and local district officials.

Although I had decided to immerse myself in village life in order to obtain the appropriate perspective on the outside world (with the ideal of Cezanne's immersion in perspective in mind), villagers did maintain more extensive contacts with the outside world than originally anticipated.[6] I was thus able to make several brief trips to Communist Party of Thailand (CPT) areas, the Hmong refugee camps along the Lao border, and distant Hmong settlements. I was also in a good position to interview the many Hmong visitors, from all parts of Thailand, who came to the village, generally to trade or to court. It was instructive to compare their views and the stories they told, with those I had gathered in the village.

Since important contacts were maintained between the village and the town of Chiangmai, I attempted to clarify the nature of these contacts and whether a process of 'urbanization' could be said to be taking place for the Hmong, through distributing a questionnaire among all the Hmong living in Chiangmai, and a separate one in a village very close to Chiangmai which has become something of a tourist resort, and where the villagers depend largely on Chiangmai for their subsistence.

I found that through concentrating on a single location, I was able to construct what I believe to be an accurate picture of many other types of Hmong settlement, simply through following the relationships which those villagers maintained with other settlements, and isolating the alternatives which defined the village. It is in this sense that I believe the village studied may justly be called an average or typical one of those in Thailand.

I had, in a less formal way than my later admission to clan membership, become a part of one of the families in the focal village. This enabled me to witness the daily routines of existence on a long-term basis much as a Hmong household member would experience them. I made many friends outside the family in the focal and larger settlements since I was inevitably less affected by the divisions of clan membership. I was able, towards the end of my stay, to repay much of the hospitality I had received through my assistance for a returned member of the CPT. I worked closely with my research assistant, to whom a great debt of thanks is also due, although in general I was the one who obtained the data in the field, and he the one who transcribed them at home. Thus I was never an 'armchair' anthropologist.

The major technique employed was, therefore, traditional participant observation, together with the use of questionnaires and recording, and the collection of biographical and case-study material. However, I was surprised to learn of another technique which I did unconsciously employ myself—that of 'shadowing' key informants, a sort of sleuthing which I undertook quite extensively. Thus I might arise before dawn, go before breakfast with the head of the household to a soul-calling ceremony for a three-day-old child which he was conducting, accompany him to the forge where tools were sharpened, visit the fields for the remainder of the day, wash with him at dusk, spend the evening by his side at a village meeting discussing a projected new water-pump, and then socialize with his sons in the village square as they wooed girl-friends, asking questions and recording where necessary. On other occasions, I was engaged in analysing texts or conducting formal interviews on genealogies and incomes, but this could rarely be scheduled in advance, since at any time a dispute might arise, or a bride be captured from a neighbouring village. I had to be constantly prepared to discard my own timetable in accordance with the actual course of events in the village.

Upon my return from fieldwork, I had the opportunity to conduct two brief field-trips to the Hmong refugees in France, and two to those in the United States, one for the purpose of attending a conference specifically about the Hmong, at the University of Minnesota. In the United States I was able to visit all the major regions of Hmong settlement and interview refugees quite extensively over a six-week period. Since this book is primarily a study of Hmong assimilation to the membership of other cultural categories, some of this data has been included in the final chapter. I have also been able to meet with and discuss my findings with a number of French researchers, and this, too, has been invaluable.

The Region

The village of Nomya was situated at an altitude of 1 400 m above sea-level in the north of Thailand, over 80 km from Chiangmai, the provincial capital (Map 1). It had been in existence for twenty-five years (its original inhabitants having migrated from near the Burmese border), and formed part of a village cluster with five other villages (Figure 1). Originally only the focal settlement, Nomya, had existed, together with the larger settlement of Hapo (the Valley of the Mulberries), and a village inhabited by members of the Xyooj clan nearby (Nomya III). Dissidents from the three hamlets had founded a new village which had entirely broken up owing to a succession of misfortunes recounted in Chapter 8. Its inhabitants had founded three new settlements along clan lines: Nomya IV, located close

Map 1   Northern Thailand

FIGURE 1
The Survey Site

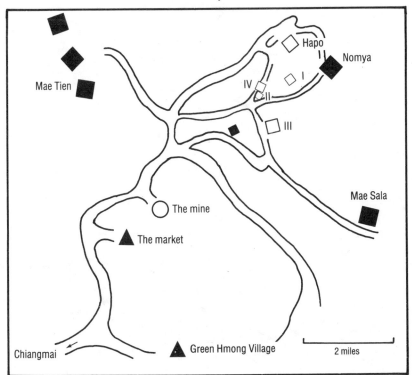

to the larger settlement from which it was joined by dissident members after a dispute over water scarcity, and Nomya I and II. Then the mother village, Nomya III, had broken up as it had grown too large for its location and its members dispersed, some moving into the focal settlement (Figure 2). Four houses of the mother village still remained. The village of Hapo was the largest village in the cluster at the time of study, with forty-six households, at an altitude of 1 350 m. Nomya itself contained only twenty-seven households, with a population of 207, and the other villages were of comparable size. All villages were situated within an hour's walking distance from each other.

Much of Northern Thailand is mountainous or hill country, lowlands accounting for only about 10 per cent of the total surface area, with limestone peaks ranging up to 2 600 m above sea-level at the highest, drained by four major river valleys which form a 'dendritic' system of tributary river-streams of the great Chao Phraya River (Keen 1973; Pendleton 1962). A technical distinction between the 'highlands' (classified as from 500 to 2 500 m and covering 60 per cent of the area) and the 'uplands' (covering 30 per cent of the area above that height) is in usage which I have not employed here, although I have specified slope and altitude where necessary. Surveys within the 'highland' area have shown slopes of under 3 per cent occupying only 1 per cent of the land, with slopes of up to 35 per cent ac-

FIGURE 2
Relocations within the Survey Site

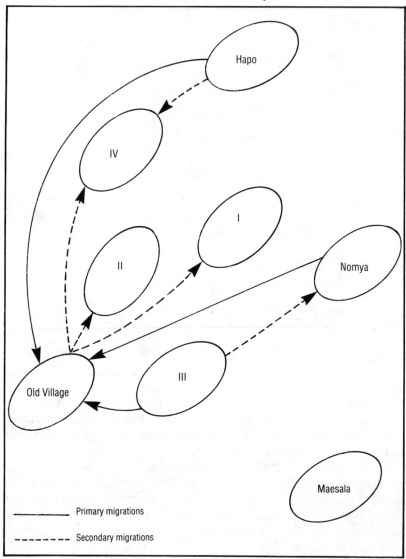

counting for a further 5 per cent, and the remaining 94 per cent of the surface area of much steeper elevations. In effect, this means that much of the land is uncultivable, with loose and thin soils on which only cereals, such as varieties of rice which do not require artificial irrigation, can be grown up to about 1 000 m above sea-level, and maize (Indian corn) above that height (besides certain beans, leafy vegetables, gourds, root crops, and some fruits). Deciduous forests give way to evergreens at approximately 1 005 m above sea-level.

Rocks in the region mostly derive from hard gneiss or schists, with areas of granite, sandstone, and the limestone outcroppings best for cultivation of the opium poppy. The major soils are reddish-brown lateritic loams, easily erodible red-yellow podzoilics, and fertile alluvials near river-beds in small quantities. There are three clear seasons, a rainy season lasting approximately from May to November, a cool season from December to February, and the hot season from February to April. The rainfall is high, with considerable annual variation, averaging from 1 300 mm in the lowlands to 2 000 mm or more near the summits of the mountains. Temperatures range from 2 °C to 40 °C, alleviated in the hills by cooling breezes with occasional frosts.

The traditional political body in the area was the valley-based chiefly principality, or *muang*, controlled by a ruler who claimed loose allegiance over the population of the surrounding hill areas (Condominas 1978). In its ethnic composition and topography, the area is akin to the neighbouring Shan States of Burma documented by Leach (1954) and the northern parts of Laos and Vietnam, bordering on to the mountainous gorges of Yunnan in southern China (Map 2). The Shan, Lao, and Northern Thai (also known as Khonmuang or Yuan) are all Tai-speaking peoples, who first established a series of Buddhist principalities in the region after a slow process of penetration and settlement, supposedly originating from Yunnan, was accelerated by Kublai Khan's sack of Pagan in Burma during the thirteenth century. Through a series of reforms between 1874 and 1901, the north of Thailand was politically and administratively united with the remainder of Thailand, with its capital at Bangkok. However, the Khonmuang maintain many differences of dialect and custom from the central Thai even today (Turton 1976), and in the days before the Second World War, when the British maintained a consulate at Chiangmai, the whole area was known as 'Western Laos', and earlier as the 'Southern Shan States'.

Like Yunnan as well as Burma, the area is notorious for its ethnic heterogeneity. The predominant population (nearly 60 per cent of the minority population of the highlands) of the hills is composed by the Karen, who also form one of the largest ethnic minorities in Burma, where they maintain autonomous areas in their own states of Kawthoolei and Kayah, supported by the armed independence movements of the Karen National Defence Organization (KNDO) and the Karenni Army (KA). During the wars between the Yuan and the Burmese, which characterized much of their mutual history between the sixteenth and eighteenth centuries, invading conquerors would remove large numbers of the conquered populations to be resettled within their own domains, since until relatively recently the ratio of population density to land was very small, and political might depended on the amount of labour a ruler could command. Chiangmai was, for example, virtually abandoned between 1776 and 1796, and the present-day Khonmuang are mixed with Shan, Lue, Thai Khoen, and Yong (Kraisri 1965). In addition to this, during and probably before the nineteenth century, an ongoing slave trade, particularly in women and

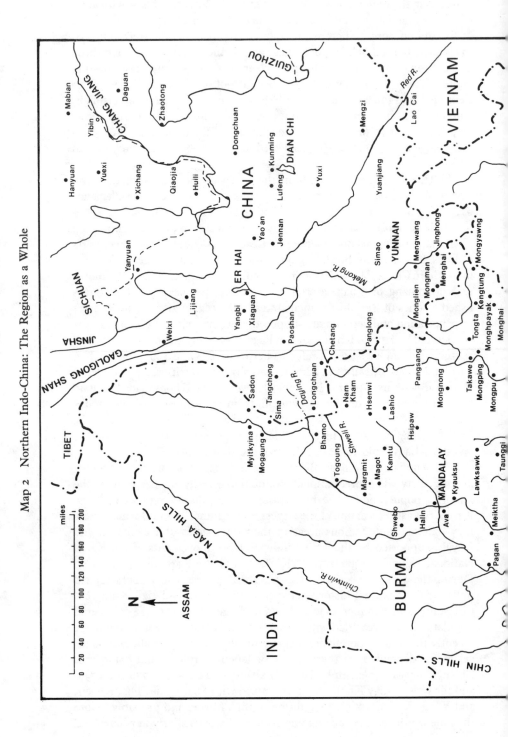

Map 2   Northern Indo-China: The Region as a Whole

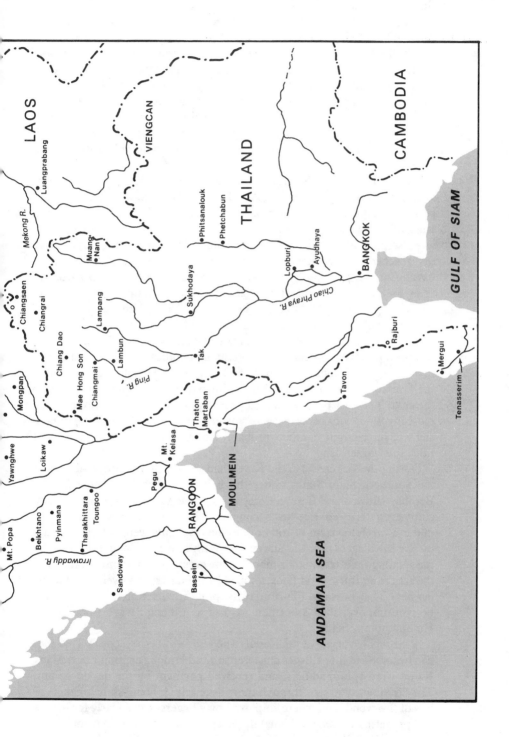

children, extended across the border regions, in which the Karen played a leading part, buying and selling captured Burmese to the Yuan, and vice versa. Thus, today, the Khonmuang population of Northern Thailand is largely heterogeneous, and this has been increased in recent years by immigrants from Laos, Burma, and China, seeking better economic opportunities in Thailand, and refuge from the conflicts in their various countries. The hills, in particular, are occupied by the members of many different minority ethnic groups, who in general maintain their separate villages. These include Tibeto-Burman groups who also frequent Burma, such as the Akha, Lahu, and Lisu, and the Mien (Yao) who are related to the Hmong.

The area in which Nomya was situated was predominantly a Karen area, settled over 200 years ago by Karen of the Skaw or Bghakenyaw group, and the village was surrounded by Karen villages with which the villagers, in general, maintained friendly trading contacts. The survey area was also populated by Lue, Lua', and Shan. Hmong scouting parties had first reached the area before the last major historical event in the area, the passing of the Japanese (whom some informants remembered bartering rifles for rice), in the mid-1930s, at about the same time that the Thais began to penetrate it. After the Japanese had left, Hmong returned in greater numbers to farm, and had slowly established settlements in the region. Some intermarriage had taken place, and some Karen children had been purchased or adopted into Hmong families. There were also conflicts between Hmong and Karen, and disputes over rights to land which the Hmong, however, blamed on the introduction of Northern Thai into the area. These disputes remained on an individual basis, however, at the most involving villages or parts of villages, and never attained the dimensions of inter-ethnic conflict, probably because of the lack of permanent titles to swidden land.

During the last century, the Karen had worked in the timber forests of the Southern Shan States under the British, and in Chiangmai for the Bombay–Burma and Borneo Companies, as elephant drivers and loggers, cultivating terraced rice-fields coupled with upland dry-rice swiddens under a careful system of rotation. Some elephants were still in local use by Karen villagers near the survey site. With growing demands for land by the Hmong, and from other more recent poppy-growing immigrants into the area from Burma and Laos, swelled by the upwards movement of many landless Thai peasants (Turton 1976, p. 122), the Karen had gradually become impoverished and often worked as itinerant wage labourers on Hmong opium fields.

Some whole villages of the Karen (who number over 150,000 in Thailand) had become satellite villages clustered around Hmong communities. Many Karen were opium addicts, and received payment by the day in opium. They frequently passed through the village selling home-spun blankets or bits of fire-wood, and where employed by villagers (particularly for hoeing and preparing opium swiddens) they slept in their houses. One murder of a Hmong by a Karen took place while I was in the village, after an argument involving a small *pob*, or lump, of opium. No animosity, however, was shown towards the Karen as a whole on account of this incident, with whom the villagers claimed they had lived in perfect understanding until the introduction of roads to the village. These had brought strife and disharmony. Karen stories, however, blamed the Hmong for employing their opium addicts and so gaining control of their lands, and rumoured that they were rich enough to bribe local officials. There were two dirt and laterite roads, loose surface until half-way, leading to Chiangmai. These were built ten years previously, after foreign missionary pressure on behalf of the Christianized Karen of the region who, they said, desired more adequate contacts with the lowlands. The roads were constructed under Thailand's United States-assisted Accelerated Rural Development (ARD) programme. One villager in Nomya, the focal village, had invested in a Toyota pick-up truck which regularly brought vegetables to the markets and the sick to the hospitals of Chiangmai.

Further down the valley a Chinese trading post, a small market town, had been established sixteen years previously, at an altitude of 1 070 m, mostly composed of Chinese merchants and traders from Chiangmai. They were attracted by the first mining operations which had begun some years before, and had married local Karen wives and set up stores dealing in every kind of goods from salt, dried fish, and tobacco, to torches and cooking pots. Most of these Chinese, and some Khonmuang, were involved in the opium business, loaning money and extending credit repayable in opium. A smaller Chinese community was settled in the nearby Karen village of Mae Sala, west of Nomya, composed of individuals who had migrated from Burma within the previous decade. Traditionally, Yunnanese traders have been the middlemen in the highland and upland areas, and formerly maintained extensive pony caravan routes between Tibet and Chiangmai, across Burma and the northern parts of Laos and Vietnam, which to some extent still exist today (Hill 1982; Forbes 1987). Although no Chinese were settled in the focal village, in the neighbouring larger settlement three (two of them Muslim) maintained shophouses. After the opium harvest, individual Chinese and Khonmuang traders set up temporary shops in each of the villages in the survey site, ostensibly selling noodles or dentistry.

In the district there were three tin-mining concessions, one government-owned (offering 60 baht a day to labourers), and two privately owned, one by a Japanese consortium (offering 40 baht a day). Many of the labourers in the mining company nearest to the village were local Khonmuang or Karen, while also including many, technically illegal, Shan, Chinese, and Burmese immigrants from Burma. No Hmong from the survey site, however, were found working in the mining concessions, owing to strong traditional sanctions against working as hired labourers for others. A neighbouring Hmong village had been moved away from land granted to one of the mining companies.

Although the villages in the cluster fell under the jurisdiction of three separate local administrative districts, no Thai administrative presence was maintained in any of the villages. However, a Hmong had been appointed headman of Hapo, and various attempts were made to select a

headman to represent the different Nomya villages. These attempts failed
because of inter-clan conflict (the Xyooj clan refused to attend a meeting
called by the local teacher, and the Thai authorities postponed their visit
until after the rainy season because the roads were impassable during the
rainy season).

A government agricultural agency maintained a post in the larger settlement, while Nomya was influenced by another project stationed beyond
the survey site. I describe these in Chapter 2. Schools had been set up in
both Hapo and Nomya besides one of the other villages (Nomya I), under
a five-year (1980–5) USAID-sponsored pilot scheme to bring Thai literacy
to the hillpeople, which fell under the authority of the Thai Ministry of
Education, and replaced an earlier Border Patrol Police (BPP) project.
These schools, for young children only, were run on a flexible basis designed to be in accordance with the labour demands of the cultivation
cycle, and were staffed by volunteers who were themselves often members
of ethnic minorities. Thus the teacher in the neighbouring village was
Karen, while the teacher in Nomya was Mien. A deliberate policy of not
employing members of the same ethnic group as those they taught was
employed in order to avoid the possibilities of the source language becoming the medium of instruction.

While reactions to these teachers were generally tolerant, Thai language
acquisition being a recognized economic asset, feelings of suspicion and
resentment were also harboured towards them, which I described later.
Reactions were more favourable towards the educational supervisor for the
whole area, who was himself a Hmong and settled in Hapo, although of a
different cultural division from that of the villagers (a Green Hmong rather
than a White Hmong). Despite such influences, it was remarkable to what
an extent the village retained its conceptual autonomy. Authority tended to
be invested in a Hmong past rather than in the Thai bureaucracy, while
a real sense of affinity was maintained with other swidden cultivators,
although they might be members of different ethnic groups, as against
salaried Thai officials or government employees.

The Hmong

The Hmong, like other ethnic minorities in Thailand such as the Mien
(Yao), Lisu, Lahu, and Akha, have traditionally favoured high altitudes of
over 1 000 m where a mixed and 'semi-pastoral' (Savina 1930, p. 214)
economy based on dry (unirrigated) rice, maize, and poppy can best be
practised, together with animal husbandry, some cultivation of fruit trees,
hunting, and gathering. This contrasts with the more lowland-dwelling
Shan, Lua', and Karen, who have tended to settle in the more intermediate
altitudes of the high river valleys, where permanent-field wet rice agriculture, often terraced, is more easily practised. Many varieties of gourds,
leafy vegetables, beans, and root crops are also cultivated in the Hmong
economy, often inter-cropped with the major crops. Other cereals, such as
millet or sesame, may also be grown in small plots, and many villages still
cultivate some Indian hemp for spinning into cloth.

While maize is grown mainly as animal fodder, for the pigs and chickens ✓ which each household rears and owns separately, rice forms the staple diet. Unirrigated upland crops have to be cultivated on a shifting basis, with fields abandoned once yields begin to decline, when a new patch of forest must be cleared and burned off before it can be planted with seed. As cultivations extend in distance from the villagers who clear them, villages fragment and villagers move off to new areas to establish new settlements. With growing overpopulation in the hills, owing both to immigration from neighbouring countries and the migrations of landless Thai peasants in search of new land to swidden, as well as to government restrictions on traditional village mobility, soil fertility has been greatly reduced in some areas and many villages have remained in the same locations for more than two decades. As primary forest has diminished, wildlife has become scarce and hunting has become less frequent, although it still continues, with some deer, pheasant, monkeys, and ant-eaters.

Although there have probably always been deficiencies of rice production in the uplands in the past, and for many centuries the inhabitants of the mountains have relied partly on crops such as tea or poppy for exchange in return for rice from the fertile lowlands, rice yields in Thailand have now fallen even beneath their traditional levels. Together with the insatiable demands of the market it supplies, this has brought the opium poppy, which survives better on poorer soils for longer periods (Geddes 1976, p. 130) into greater demand as a cash crop. Opium, laboriously tapped from the poppy heads, is not only exchanged directly or indirectly for rice from the plains and fertilizer to maintain crop yields, but is also used extensively as the very effective medicine it is.

Houses are built on the ground, unlike the Thai pile houses, out of upright wooden shingles usually notched together or bound with hemp rope and creepers, and thatched with teak leaves or cogon grass. Richer households may invest in zinc or polystyrene roofing; poorer families may construct their houses entirely out of pieces of split bamboo. Villages may number between seven or fifty households, often arranged in a horse-shoe pattern just beneath the crest of a mountain, sheltered by a belt of forest and close to a source of water. Water may be piped down the mountain to the village through a raised semi-trough formed out of split bamboo, propped up at intervals by a series of forked sticks, and is collected mostly by the women in wooden or metal buckets. Tall clumps of cooling bamboo, banana or peach plantations are usually maintained near the village, while herbal gardens flank the neighbouring slopes. Around the houses, rough wooden granaries are constructed, raised from the ground to protect against field mice and scavengers where grain is stored. Shelters are put up for tethered horses and pack ponies, and small chicken-coops built to which a multitude of fowls return at night. The village is cleared of refuse by the rooting pigs, which are not usually penned. Semi-wild mongrels attach themselves to the porches of each house, which they guard and from which they are thrown occasional scraps of food.

Hmong began migrating into Thailand from Laos after 1850, emanating from the Guizhou–Sichuan–Yunnan region of south-western China. Together with other groups, the Hmong are sometimes termed 'aboriginal Chinese' because of the likelihood that they were among the very first groups in ancient, pre-feudal China. As the northern Han Chinese penetrated the southern regions (Wiens 1970; Fitzgerald 1972; Moseley 1973) colonizing Guizhou in 1413, a series of pacification measures and land campaigns led to the great 'Miao' rebellions of 1733–7 and 1795–1806 (Lombard-Salmon 1972), the bloody repressions of which drove many Hmong out of China in two successive waves, during 1800 (when large numbers also arrived in Yunnan) and 1860, into Tongkin (north Indo-China) and the northern Lao states.[7] More left after the last great Miao rebellion in Guizhou of 1855–72, synchronized with the Tai Ping revolt. Some 5,030,897 still remain in China, where they form the fifth largest national minority, together with the (linguistically cognate) Hmub and Qob Xyoob associated, respectively, with the first two rebellions above, centred in Guizhou, Yunnan, Sichuan, and Guangxi.

The Hmong of the fieldwork area were mostly descendants of the earliest settlers in Thailand, who had entered near Chiang Khong on the Lao border and followed the range of mountains along the Burmese border down into the Thai provinces of Maehongsorn and Chiangmai. Some of the eldest, however, had had parents who had claimed to have lived in China. In Thailand, the total number of Hmong is estimated at a maximum of 80,000 (cf. Table 1), about equally divided between the two main cultural divisions of the Hmong in Thailand, the Green and the White, who maintain strong distinctions of custom, costume, and dialect despite the increasing frequency of intermarriage between the two (Cooper 1984, p. 51), and normally maintain separate villages.[8] Miao groups have also been reported in Burma (Scott 1900, p. 597; Scherman 1915; Binney 1971, p. 348), and this was confirmed by Hmong in the survey area, while Davies (1909, p. 131) chronicled Miao villages as far north as the Tibetan state of Mi-li.

Hmong culture is rich, complex, and infinitely varied. Many elements

TABLE I
Hmong Population in Thailand

Province	Population	Households	Villages	Year Surveyed
Kamphaengphet	1,451	162	7	1977
Chiangrai	7,875	1,229	22	1976
Chiangmai	9,007	1,202	54	1979
Tak	7,072	1,029	20	1977
Nan	6,205	684	14	1979
Petchaboun	8,632	326	2	1977
Phrae	496	62	3	1975
Maehongsorn	1,836	230	16	1977
Lampang	576	73	3	1975
Sukhothai	269	41	1	1977
Total	43,419	5,038	142	

Source: Tribal Data Project, Tribal Research Centre, Chiangmai.

are shared with the Chinese, and it is towards an exploration of the significance of this that the present work is directed. Their society is patrilineally
based, and divided into various clan-like, exogamous surname groups
and descent groups within them distinguished by differences of ritual,
especially during the lengthy and involved funeral rites. A bilateral form
of preferential cross-cousin marriage survives in the form of child be-
trothals, polygyny is permitted, and the levirate practised regarding elder
brother's widows.

The *xeem*, or clan, truly forms the basis of all social interaction, and
individuals are primarily classed according to their membership in a particular clan. Weddings between clans are celebrated with great pomp and
expense, but the only usual occasion on which large groups of Hmong
meet together on an inter-clan basis is at the New Year ceremonies, held
for a minimum of three days after the rice harvest has been brought home.

Probably on account of the lack of permanent titles to land, as Lemoine
(1972a, p. 192) points out, there are no formally specialized roles in Hmong
society, though individuals may specialize as blacksmiths, wedding go-
betweens, or shamans. Shamans may be male or female, and shamanism is
to some extent hereditary (see Chapter 5). The shaman specializes in re-
calling the wandering soul whose absence from the body brings sickness
and, in cases of long separation, death.

In Thailand, the Hmong have borne the brunt of the anti-opium campaign launched by the Thai government with international assistance from
the United States Operations Mission (USOM) since 1959, and the United
Nations (UN) since 1971. In the mid-1960s, a number of Hmong, alienated
by bureaucratic policies towards them, joined forces with the CPT, and the
popular image of them today in Thailand remains one of opium-producing
'komunit' (the Thai colloquialism for 'communist'). This stereotype is
contradicted, however, by the large numbers (over a third of the 300,000
calculated to have lived in Laos) of rightist Hmong who, since the ending
of the Indo-China Wars in 1975, fled as refugees into Thailand from Laos
after the Pathet Lao gained control of the country. These Hmong refugees
are, in general, carefully segregated by the Thai authorities from the
Hmong of Thailand although, as we shall see, contacts do take place.

A great number of the former have now been resettled in third countries
such as France, the United States, Australia, and even Argentina and
French Guyana, but over 50,000 remained in refugee camps within Thailand
at the time of fieldwork. A number of these were employed as 'resistance
forces' engaged against the Vietnamese-supported Pathet Lao government
in Laos, and Hmong resistance movements, under the leadership of various
factions, have survived in Laos.

Under the pressure of various government anti-narcotic projects spear-
headed by the UN/Thai Project since 1971, several Hmong and other
upland villages in Thailand have now surrendered poppy cultivation and
turned instead to wet-rice or mixed fruit and vegetable agriculture of
a permanent-field nature. Often these villages, however, have become
economically unviable in the process and have been transformed into
'model' villages demonstrating the success of official agrarian policies for

the benefit of visiting dignitaries. A recent trend has been to turn these villages into tourist resorts for Thai and foreign visitors. A particular example of this was the Mae Sa Mai valley, very close to Chiangmai, the site of a United Nations Development Programme (UNDP) pilot scheme. Often, however, the continued presence of opium addicts within these villages necessitates importation of opium from other upland villages. Even the refugee camps require opium. Thailand's booming tourist trade, at the time of my fieldwork the country's largest foreign currency earner, has turned Chiangmai town into a centre of 'hilltribe handicrafts' produce. Traditional Hmong costumes, with their strikingly beautiful appliqué, batik, and embroidery work, now fetch good prices on the tourist market, and several shops in Chiangmai specialize in the selling of such articles.

Hmong were employed as entertainers performing 'song-and-dance' routines in two cultural shows aimed at presenting (a certain kind of) Northern Thai life to the tourists in Chiangmai. Although the commercialization of the handicrafts industry had not yet reached Nomya or any of the villages in the survey site, other contacts with Chiangmai were maintained. Several Hmong from the district had sent their sons to be educated there—one from Nomya, and two from Hapo, while seven children from Hapo attended an ex-BPP primary school at the nearby Chinese trading post. A government radio broadcasts in Hmong twice a day from Chiangmai. Rice was bought from Chiangmai to supplement insufficient yields from the dry-rice fields, while some innovative products, such as lettuce and broccoli, were taken there to be sold in the markets. Extensive contacts for medical purposes also took place.

The Language

Linguistically, the term 'Hmong' (or 'Mong' since it is unaspirated by the Green Hmong) refers to the western branch of the Miao language family, which also includes the Hmub and Qob Xyoob, which terms are cognate with 'Hmong' (Downer 1967; Lemoine 1972a). The latter two were referred to, respectively, by the Chinese as the Hei (Black) Miao, and the Hung (Red) Miao, while the former included the Hmoob Dawb (White Hmong), Hmoob Ntsuab (Green Hmong), Hmoob Quas Npab (the 'Striped Armband' Hmong, small groups of whom now live in Thailand, speaking a variety of White Hmong, and who may be part of the group known to the Chinese as the Hua or 'Flowery' Miao), and the Hmoob 'Ntsu' (Ruey Yih-fu 1960), known to the Chinese as Yob Tshuab ('Magpie') Miao.

Miao is related at its upper levels to the Yao dialects in a language family known as Miao-Yao, for which a proto-Miao-Yao can be constructed, but no relationships to any other languages have yet been established, although the whole group has been strongly influenced by Chinese. The Miao-Yao languages are probably to be classed as Sino-Tibetan, although scholars are in disagreement about this (see, for example, Benedict 1975). Hmong has eight tones: high, high falling, mid, mid rising, low, low heavily gutturalized, lowest glottally stopped, and (rarely) low rising (indicated by the following roman end consonants: -b, -j, -, -v, -s, -g, -m, -d).

As Matisoff (1983) has recently put it, '. . . it is Miao-Yao which is the tonal champion among the tone-prone languages of East Asia'. He adds that the Miao-Yao languages have 'extremely complex phonologies, with elaborate tone systems, pre-nasalized consonants, pre-glottalized sonorants, post-velar stops, dentally, palatally, and laterally released affricatives, voiceless nasals, central and back unrounded vowels. . . . Such features are also to be found in the other language families of the area, but it is as if Miao-Yao had developed them all to the *n*th degree' (Matisoff 1983).

Nomya Village

The twenty-seven households of Nomya include one belonging to the Mien schoolteacher, and one belonging to a Khmu', a member of another mountainous ethnic minority originating from Laos, who had married a Hmong wife.[9] The remaining twenty-five were distributed unequally between three patrilineal, exogamous clans or surname groups—the Vaj, Yaj, and Xyooj (Figure 3). All the families were interrelated through a variety of affinal ties centring on household number 12, whose head constituted, therefore, the grand old man of the village (Figure 4; see also Chapter 5 and Appendix 2).

The village had been established by him (Xeev Hwm) and his brother, of the Yaj clan, twenty-five years previously. Two years later, the current headman of the village, who was of the Vaj clan, had married one of Xeev Hwm's daughters and moved into the village, forming the nucleus of the Vaj clan households there. Eight years after that, the Khmu' had married the daughter of the younger brother of the old man of the village, and had moved into the village, but with a very different status to that which was to be accorded to the old man's Vaj clan son-in-law, Vaj Xeeb.

Political power was vested in the village's acting headman, Vaj Xeeb. Vaj Xeeb was said to have achieved his position largely through marriage to one of Xeev Hwm's daughters, since he had previously been poor in relation to his father-in-law and had married into the village matrilocally, while over the years this situation had been reversed. Xeev Hwm is, in fact, on record as saying that only through marriage to members of his own clan (the Yaj), could members of the Vaj clan achieve anything at all. Unless Vaj married Yaj, he said, they would always be poor and suffer untimely deaths. There was thus a good deal of opposition and rivalry between these clans, despite the strength of their affinal interrelationships. The Vaj, for instance, considered the Yaj as untrustworthy and impossible to deal with, and the Xyooj minority as only slightly better.

The Xyooj families had moved into the village in a piecemeal fashion over the previous five to seven years, mainly for reasons of illness and misfortune in their previous village (see Chapter 8). The search for land was the most frequent answer given to questions about the reasons for relocation, especially by the earlier pioneers. Later settlers tended to accord primary importance to the need to be near *kwvtij* (all clan relatives, referred to by a combination of the words for older and younger brothers). Sickness and

FIGURE 3
Household Composition of the Village

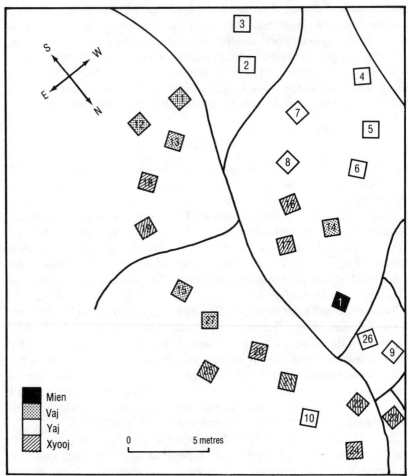

Key

Household Number	Name	Clan	Household Number	Name	Clan
1	Laopa	Mien (Yao)	15	Tsav Xwm	Yaj
2	Vaj Xeeb	Vaj	16	Vam Nrhuag	Xyooj
3	Suav Yeeb	Vaj	17	Vam Pov	Xyooj
4	Yis	Vaj	18	Nom Tuam	Xyooj
5	Tsuj Ntxawg	Vaj	19	Lwm	Xyooj
6	Nkaj Suav	Vaj	20	Caiv Phab	Xyooj
7	Nkaj Huas	Vaj	21	Vaj Neeb	Xyooj
8	Ntxhoo Pov	Vaj	22	Tooj	Xyooj
9	Neeb	Vaj	23	Pob Zeb	Xyooj
10	Nyiaj Paiv	Vaj	24	Tsuj Xwm	Xyooj
11	Siv Yis	Yaj	25	Khav	Khmu'
12	Xeev Hwm	Yaj	26	Tsheej	Vaj (new arrival)
13	Ntxhoo Xab	Yaj	27	Kaub	Yaj (new arrival)
14	Looj	Yaj			

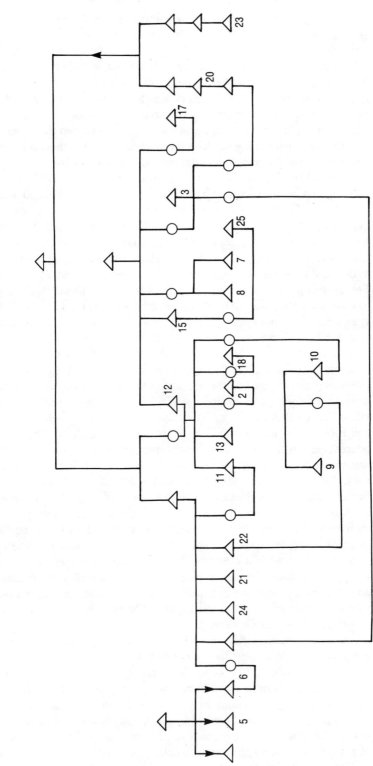

FIGURE 4

Simplified Kinship Diagram of the Major Relationships in the Focal Village

misfortune, however, in the original settlement, was a frequent response among both kinds of settlers, often associated with other reasons for relocation.

While in terms of population, the Vaj, with 92, greatly outnumbered the Yaj, with 47, they outnumbered the Xyooj to a lesser extent, who with 68 in fact outnumbered the Yaj (see Appendix 2). Thus the Vaj did not have an outright majority in the village, and it was in their interests to remain on good terms with the small Yaj group, who had originally founded the village. It was only through maintaining ties with the dominant Vaj that the Yaj managed to retain any power in the village, since they were in the minority regarding both the Vaj and the Xyooj. The Xyooj, however, were excluded from effective decision-making in the village by the alliance between the Vaj and Yaj. Thus the balance of power in the village was an uneasy one, and rested on a trium-virate formed by Vaj Xeeb, the headman, Xeev Hwm, his father-in-law, who had founded the village, and another old man of the Vaj clan, Suav Yeeb, who lived next door to Vaj Xeeb and whose eldest son was the only person in the village said to be clever enough to become a serious rival to Vaj Xeeb for the headmanship. Although Suav Yeeb's family had moved into the village twelve years after Vaj Xeeb, the older man was the ritual head of the Vaj clan descent group of which Vaj Xeeb was a member. It was Xeev Hwm's younger brother whose daughter had married the Khmu'.

Although considerable dislike and disapproval of this liaison was voiced by members of Xeev Hwm's clan, in particular by his eldest son, no oppo-sition to the marriage had taken place at the time it was made, apparently because the liaison had occurred while the Khmu' was resident in another village, or in other words, the girl had become pregnant before the Khmu' had moved into the village. The Khmu' remained, however, one of the poorest men in the village, and spent most of his time living in the fields, sometimes working for his father-in-law since no bridewealth had been paid. Along with eight other household heads, he was addicted to opium (which I calculate on the basis of smoking several pipes three to six times a day), and in addition had a large family of nine children to support. While their father retained his ethnic clan name, the children had become com-pletely Hmong in every way, and were free to marry anyone except mem-bers of their mother's clan, the Yaj, to which they were considered to belong. Uxorilocal residence is a disgrace, and will only normally be adopted when poverty forbids the payment of bridewealth, or in rare cases when only a daughter is left to look after her parents (in which case bride-wealth is also usually lowered).

Although, theoretically, the old man's younger brother should have proved a valuable support for the position of his brother and thus the whole Yaj clan in the village, in practice this was not the case, since he was slightly deranged and was in fact said to be insane (*vwm*) by the headman. He had been a practising shaman, among seven others in the village, but one day he had given it all up and thrown his shamanic equipment away. This was said by the headman to be because he had gone mad following the discovery of his wife's adultery, but an alternative explanation offered was that he had been so intensely ill and for so long that he had lost faith in shamanism and given it up in disgust.[10] As we shall see later, this case is of interest for the light it sheds on conflicts in the village between the practice of traditional shamanism and the introduction of recognizably effective modern medical techniques associated with the adoption of Christianity.

Decisions regarding village matters had to be taken on a consensual basis, once agreement had been reached between the two rival Vaj clan factions represented by the headman's household and the household of the ritual head of their clan, and between them and their affinal Yaj clan relatives. Care had to be taken, however, not to alienate the nine Xyooj clan households entirely, since in the event of a serious dispute they always had the option of moving out of the village altogether, which would have weakened the village as an administrative unit and as a unit of self-defence. They were, in fact, contemplating such a move during the period of research.

Residence in the village is divided between the house in the village and the rough field-house constructed in the fields, between which individual cultivators move, either on an overnight basis or for longer periods. Short-scale relocations take place as in the case of a Xyooj clan household which moved to a nearby village during the period of fieldwork, and a Yaj man who moved into the village after marrying a Xyooj girl. Major relocations of villages or village segments of the type pondered by the Xyooj households also take place, although they are now severely restricted by government limitations on traditional mobility, and, as a consequence, individual migrations are more common. These kinds of movement should be compared with the long-term migration of the Hmong, over a period of thousands of years, from a country possibly to the north of China into China, Indo-China, and recently the West, and the current movement of rural Hmong into urban occupations in Chiangmai, represented in the field site by the youngest son of the old man of the village, who was at school in Chiangmai town.[11]

1. All references specifically to the Hmong in the following pages are to the White Hmong, and not necessarily to the Green Hmong. Where more general reference is made, for example, to 'the position of the Hmong in Thailand', both groups are included. In referring to China, unless the group is specified, I have used 'Miao' to include also 'Hmong'. Where following secondary sources, I have used their terms. My own data, however, refer specifically and exclusively to the Hmoob Dawb (White Hmong) and not to the Hmoob Ntsuab (Green Hmong, often known erroneously as Blue Hmong). On the Green Hmong, see Lemoine (1972a).

2. On this 'observer-effect', see Cassirer 1953 (in Firth 1964, p. 41).

3. Preface to *The Lyrical Ballads* (1798).

4. Cooper (1984, p. 74) seems to do this for example, but may have been misled by Lemoine, 'un lignage est aussi un village' (1972a, p. 188). In my experience, the *zos* always referred to a *residential* group.

5. For this I am most grateful for the individual tuition of Professor Gordon Downer of the University of Leeds.

6. Merleau-Ponty (1947) describes how the apparent distortion of Cezanne's work arose from his attempt to place himself in the centre of what he was painting.

7. 'Miao' is a derogatory term applied to other groups as well as the Hmong in China, and its derivative in Thai is 'Meo'. Where historical sources refer only to 'Miao', or where the languages of which Hmong is one are referred to, I have retained the usage.

8. This is Bertrais's (1978) estimate, which I would agree with, since it is undoubtable that government surveys have in the past not covered many areas of the hills.

9. See Izikowitz (1951) for the closely related Lamet.

10. Adultery can only take place where a married woman is involved, since polygyny and pre-marital sex are permissible. This means, in practice, that only women can commit adultery.

11. I have not included the real names of informants in this book except where informants specifically requested their names to be used, for example, in the recording of ritual texts.

PART II

2

The Economic Dilemma

Introduction

WHEN one has lived for eighteen months in a village whose remoteness from the state of which it was a part allowed its inhabitants to regard it as the outpost of a Chinese civilization long since dead and vanished, it becomes difficult to revivify that experience for others. It was partly the village's physical remoteness and inaccessibility from the main power-centres of the present-day state of Thailand which enabled such an idealized view of it to be maintained, partly, too, its ecological affinities with the mountainous hinterlands of south-west China and the heavily Sinicized Shan States of what is today Burma. The sense of belonging to a wider community, although that community might no longer exist, was an all-pervasive one. Yet the distance of the village from the state was, nonetheless, largely a conceptual one. For the presence of roads, fertilizers, sewing-machines, cars, radios, and aspirin, should have long since eroded the village's sense of its own autonomy. That it had not done so was a tribute to the strength, power, and vitality of the conceptual system which guided and ordered villagers' lives. It was the village's very transitionality which struck me as perhaps its most remarkable feature—and yet one which was not unique, but shared with many Third World ambiances, where progress and tradition jostle uncertainly in an uneven development, and some parts of the social structure, it is clear, change much more slowly than other parts, or not at all (P. Anderson 1974, p. 18). But more remarkable, perhaps, in the context of a purely oral tradition, which was, above all, what imparted such resilience to the conceptual system.

At night the mountains are alive with crickets, stars, and wind. By day the women sit outside the porches of the houses they guard, embroidering clothes for their children and menfolk. The shaman begins to tremble before the altar in the house of the sick as he goes into trance, he chants in a high rhythmic voice, conjuring up his legions of supporting forces as his body mimes the violent motions of a rider on his horse. The men are at the fields with their families, hacking brush, planting corn, and catapulting small birds to cook for the midday meal. On the wide crests of the mountains where fields of variegated poppy stretch towards the valleys, a girl lies in the cool sunlight by the edge of her family's field, embroidering idly as she watches over her baby brother, practising a love song for the lover she will meet after dark, after supper, in the forest. Such a life had endured for hundreds of years.

While harkening back, through the weaving and embroidery of the elaborate clothes of childbirth and of death, through the pomp of the wedding ceremony and the mysteries of shamanism, to a great tradition which still regulated everyday conduct, needs and wants remained of an astonishing simplicity: clothes and rice, salt, guns and bullets, an absence of taxation or externally imposed authority. Authority remained in the past, which had been invested with the majesty of the supernatural. Such investiture was significant enough to inform behaviour and regulate conduct.

It was not the tension between one stasis and another, which normally defines social change or temporality (Firth 1964), which defined this village's transitionality. Rather it was the tension between stasis and change itself. And, while changing, the village had preserved its traditional stasis. While absorbing many modern and innovative inputs, the conceptual structure which defined the village and ordered daily life there remained intact. It was this tension which made living there, and the role which was eventually found for me there, such an extraordinary experience. It was this tension, too, which gave rise to the very considerable apprehension and anxiety about the future which was evident among the villagers. The transitionality of the White Hmong culture, of which this village was a part, increased the measure of 'uncertainty'[1] which attended daily, and yearly, decision-making processes. The composition of the village itself was determined by a series of dilemmas or alternatives which had, as yet, found no final resolution. The first of these, and the most primary, was economic in nature. And it centred on whether, or not, to produce opium.

Villagers experienced considerable anxieties about this. Through constant radio propaganda and development didactic, most of them were aware of the evils which opium production caused in other parts of the world. Yet, without the production of opium, it was difficult to see how they could survive. The village was itself in Chiangmai province, in one of the major opium-producing areas of Northern Thailand. Roads had brought easier access to lowland markets for opium as for other goods, although the only road to the village remained a dirt track which was impassable during many of the monsoon months of the year (May–September). Partly owing to government restrictions on traditional village mobility, the main settlements had been in existence for over twenty years. The soil deterioration which would normally have occurred after twelve to fifteen years (Friedman 1975; Keen 1978) had been checked to some extent by the use of fertilizers donated by development projects for crops alternative to opium such as coffee and broccoli. Many fields, however, were already over a day's march from the village, and necessitated lengthy overnight stays before returning to the village. Opium remained the primary cash crop of the village, with yields averaging 7½ joi per year (Appendix 4).[2] The price of a joi varied widely with the very varied and uncertain climatic and ecological conditions attending poppy cultivation, but, in 1983, averaged 4,000 baht (Table 2). This meant that villagers could sometimes derive incomes well above the rural Thai average,[3] although the expenditure necessary for the

TABLE 2
Average Annual Prices of Opium (by the joi)

Year	Price (in baht)
1974	2,000
1975	3,000
1976	1,500
1977	1,100
1978	1,500
1979	20,000
1980	15,000
1981	3,000
1982	2,500

Note: These prices were those paid for one joi (1.6 kg) of opium after the harvest, to villagers in villages in the survey site.

purchase of rice and employment of labour cancelled this margin out. Three children from the survey site had been sent to study in Chiangmai and two of the villagers in the site owned pick-up trucks, which again facilitated opium trading. At the same time, because of the uncertainty of its production and constant government pressure, villagers would have done almost anything they could to have been able to replace opium poppy cultivation with the cultivation of some other crop. Under the influence of various government projects established since the Opium Act of 1958–9 made the production, consumption, and sale of opium illegal in Thailand, many had tried to replace poppy with other crops, with disastrous results which I detail later (see also Cooper 1979; Lee 1981).

Thai Government Policy towards the Minorities

The Thai government and its associated agencies have, in the past, maintained posts in the hills at least partly for strategic reasons. As early as 1959, when the government initiated projects for the 'development and welfare' of the hill-dwelling minorities in the North through the establishment of a National Hill Tribes Welfare Committee, the projects ran along the four lines later formulated by the Director-General of the Department of Public Welfare (PWD) responsible for the implementation of the projects, as (Suwan 1969):

1. To prevent the destruction of forest and sources of natural streams by encouraging stabilised agriculture to replace the destructive shifting cultivation practised by the hill tribes
2. To end poppy growing, by promoting other means of livelihood
3. To develop economic and social conditions of hill tribes so that they may contribute to national development, by promoting community development among the hill tribes grouped in settlements
4. To induce the hill tribes to accept the important role of helping to maintain the security of national frontiers, by instilling in them a sense of belonging and national loyalty.

Prior to that time, projects in the hills had been run by the Border Patrol Police (BPP), an organization equipped and funded through the United States Operations Mission (USOM).[4] The BPP was set up in May 1953 as an operating division of the Thai National Police to maintain security and gather intelligence in remote frontier regions, and was the first government agency to have any contact with the uplands-dwelling minorities, initiating an upland school project in 1955. However, BPP projects did not cease with the advent of other projects. From September to December of 1962, two BPP teams, aided by aerial reconnaissance, conducted an intensive survey of villages in Chiangrai and Nan provinces, establishing contacts with Hmong and Mien settlements, and selecting sites for short take-off and landing (STOL) airstrips. From each village, five youths were selected for training in Chiangmai, while a more permanent presence was maintained in 'key' villages, where schools and 'dispensaries' were constructed. Early in 1964, some 100 uplanders were trained in first-aid, agriculture, and sanitation, and received political indoctrination before returning to their villages to establish aid and information centres. This project, training hillpeople as village guards to form border security volunteer teams, was conducted in conjunction with the Communist Suppression Operations Command (CSOC).[5] It was expanded in 1968 and still continues today. Other projects conducted by the BPP in the hills have been the establishment of 'boy scouts' groups containing over 2,000 members, a mobile development unit project under which the labour forces of villagers are 'catalysed' for the construction of schools, roads, and bridges in upland areas, some medical assistance, and the sponsoring of the production of 'hilltribe artefacts' for a shop maintained by the BPP in Chiangmai. As Kunstadter (1967) points out, these projects have formed part of a co-ordinated approach. The BPP might initially persuade villagers to build an airstrip in return for food, tools, and incidental medical attention, on which a plane would eventually be landed, often bearing technical and medical personnel. Seeds and animal breeding stock might be distributed, pictures of the Thai King distributed, complaints listened to and propaganda speeches made, and thereafter the patrols would return on a monthly basis. I have dealt with this programme at some length because it illustrates the form which many later development projects have taken in Hmong settlements.[6]

Many other projects have been undertaken, such as the semi-autonomous 'King's Project for Hilltribe Development and Welfare', and projects launched under the auspices of different government departments and ministries, such as the Royal Forestry Department's huge reafforestation programme for large areas of tribal land, or the Department of Public Welfare's project to establish (in 1968) a hilltribe radio station in Chiangmai which now broadcasts to all minority villages twice daily in their own languages. Over the years there has been an increasing internationalization of development programmes affecting the upland areas, particularly since the initiation of the Joint UN/Thai Programme for Drug Abuse in Thailand (UNPDAC) in 1971. Since 1980, international agencies and projects involved in educational, medical, infrastructural, and crop-substitution projects directly affecting the lives of the Hmong and other uplanders have

included the US Department of Agriculture (USDA)-supported King's
Project, the World Bank-supported Highland Agricultural and Social
Development (HASD) Project of the Australian Development Assistance
Bureau, the UN/Thai Project administered through UN Development
Programme (UNDP) offices, later succeeded by the Highland Agricultural
Marketing and Production (HAMP) project, the USAID-supported Mae
Chaem Development Project, the Thai–German Highland Development
Project, and the UNDP/Food and Agricultural Organization Mae Sa
Integrated Forest and Watershed Land-Use Project (Map 3).

However, the multiplicity of these projects and agencies, and the very real

Map 3 Development Projects in Highland North-west Thailand

Source: Highland Agricultural and Social Development Project Inception Report (1982).

rivalries between some of them (for example, between the BPP and PWD), disguises neither the extent of overlap and liaison there is between them (the Ministry of Public Health, for example, constructing with UN aid water systems and public toilets in several key PWD villages), nor the essentially strategic reasoning which has underpinned them.

In 1973, a Government White Paper on insurgency defined the following factors of the hillpeople's vulnerability to communist propaganda: (1) The primitive nature of their social structure; (2) The inaccessibility of their settlements, and (3) Their status as ethnic minorities.[7]

But it has long been the second factor which has been the target of government policy towards the uplands, since the inception of these programmes.

The PWD programme, initiated after the passing of the Anti-Opium Act of 1958, has been rationalized as following four main branches. The first of these was the Nikhom or 'Land Settlement' project, which was a continuation of earlier land settlement projects already launched in other areas of Thailand by the Bureau of Land Settlement in the Ministry of the Interior. The primary aim here was to 'persuade the hill tribes living scatteredly to move into the project areas and settle down permanently' (Suwan 1969).

To this end, strategic centres were set up at four intermediate altitudes in different provinces equipped with stores, dispensaries, schools, and supervisors, and it was expected that a mass migration of impoverished swidden agriculturalists into the areas around them would occur. Very soon it transpired that this was not going to happen. In 1966, General Prapas, in his capacity as Minister of the Interior, confessed that 'the immediate success of the Land Settlement Program had not measured up to the earliest hopes for them because the tribal people are slow to abandon their old independent ways and move into the resettlement areas' (Prapas 1966).[8]

The primary aims of stabilizing the population of the hills and gaining access to remote villages persisted. Under the second branch of the programme, 'mobile development' units, consisting ideally of an agricultural, a health, and a 'social' worker, visited selected sites for a period of a week to two weeks per month (Mandorff 1967b). It was decided to use the *nikhom* as bases from which development could be brought to the people since, it was clear, the people were not going to come to development. Yet the concept of mobile development, akin to the 'Strategic Hamlets' approach pursued by the Americans in Vietnam at the same time, was, in effect, a continuation of much earlier tactics utilized by the BPP in their approach to upland villages, and one which still characterizes most, if not all, projects in the area today. Posts are established in selected villages for experimental crop trials, Thai literacy projects for young children, or medical studies of opiate addiction, and then visited, often with great material difficulty, and often by a series of different personnel, by Thai extension workers who find themselves isolated and alone in cultural communities which are not their own, have no real or abiding interest in the villagers or their problems, and leave as soon as they feel they can. All too often, it will be found, a village 'hospital' or 'dispensary' boils down, in fact, to a Thai teacher who is resident in the village for only very short periods each month, equipped with a medicine cabinet from which he dispenses aspirin and bandages to the old women of the village.

Under the third branch of the PWD Development and Welfare Programme, a Tribal Research Centre was established in 1965 in Chiangmai town which has conducted extensive demographic and ethnographic surveys of upland groups, with assistance from the South-East Asia Treaty Organization (SEATO) and the British and Australian governments. The Centre, however, has in the past employed few permanent minority members (Lee 1981, p. 301), and thus is limited in its impact on the actual needs of the minority people.[9]

The fourth branch of the projects co-ordinated by the PWD has been a Buddhist missionary project, which is dealt with in more detail later. While new ideas of development have emerged over the course of time, such as the notion of 'zonal' development which surfaced during the early 1970s, they have betrayed a similar dependence upon strategic concerns, the same intent in stabilizing or relocating the upland hill-dwellers, as previous ones. The advantages of zonal development, which involved the concentration of people and intensification of production on lower slopes in order for the steeper slopes to revert to forestry, were defined as follows: '(1) Stabilising the people and agriculture; (2) Providing the right kind of communications;[10] and (3) Protecting and planting trees' (Keen 1973).

While not wishing to decry the very real international and governmental concern about deforestation in Northern Thailand, it is the first of these perceived advantages which has most directly affected people at the local level. As a senior Royal Forestry Department official told me, 'It is our duty to replace forest and keep these nomads fixed in one place' (7 April 1981).

In 1976, the PWD declared that while development teams would continue to be despatched to areas which had not yet been designated 'Development Zones' and would remain in areas unsuitable to be so designated, areas *abandoned* by hillpeople would be established as 'Development Zones' in which integrated development projects would be implemented for the remaining hillpeople. Such statements presumed relocation as a first premise, and entailed the seizure of the large areas of fallow land which the nature of swidden cultivation demands if swidden is to remain a viable mode of subsistence. In supporting such plans, the government, it was said, would be able to control the movements of hillpeople in and out of them, which would '(1) Prevent deforestation and opium production, and (2) Facilitate the detection of any movements which might adversely affect the political position and security of the nation.'[11]

Here the strategic intentions were quite clearly primary.

Hmong Resistance and the 1967 Revolt

Traditionally, Bangkok links with the North have been weak; Chiangmai was only gradually made part of the modern state of Thailand, and before

1874 was an autonomous Tai kingdom which had held formal tributary and ceremonial relations with its upland populations. Even today, the North has more ecological and demographic affinities with the neighbouring Shan States of Burma, the Chinese border, and Laos, than with the densely populated, humid plains of the central and southern regions of Thailand. The intensification of government-sponsored attempts to claim the hills and remote areas for the Thai state seems to have exacerbated tensions between the indigenous populations and the new settlers. For not only through establishing peripatetic and intermittent posts in the hills did the Thai presence visibly increase. Concurrently, thousands of lowland Northern Thai (Khonmuang) peasants, owing to increasing rural indebtedness and the loss of their land, migrated up into the hills to become swidden farmers, depriving the uplanders further of the high ratio of land to population required for swidden agriculture. Inevitably, these settlers were protected and to some extent attracted by the presence of government installations. This occurred to such an extent that today a large number of shifting cultivators are, in fact, ethnic Thai (Kunstadter 1983. Cf. Turton 1976; Cooper 1984):

Another serious problem is the under-counting of upland Thai people. Their number must be very large, given the estimate of at least a quarter million Northern Thais engaged in swiddening in the late 1960s, and the indication that the number of Thais dependent on swiddening is probably increasing rapidly... the total upland population in the mid-1970s was probably well over half a million people.

Increased contacts between the structures of the lowland Thai bureaucracy and the Hmong population of the hills led to a series of violent clashes and sporadic resistances. The chief among these arose from the burning to the ground of Doi Chompoo, a Hmong village in the province of Chiangrai, after various attempts by local officials to exact payment for the cultivation of poppy had been rejected by the Hmong villagers. A widespread ethnic Hmong rebellion broke out in four provinces in 1967–8, affecting large areas of Chiangrai and Nan provinces. The Royal Thai Army and Airforce were despatched to the area, and the government treated the movement, which in its origins was a local one, as a full-scale insurgency. Troop assaults, napalm, and heavy artillery strikes were employed, and hill villages suspected of harbouring insurgents were bombed from the air. Local Chinese militia and members of other minorities, such as the Akha, were also mobilized against the Hmong. As uplanders and highlanders alike fled into the forest or into the lowlands, five major refugee centres were established by 1971 in the provinces of Tak, Nan, Chiangrai, Petchaboun, and Phitsanalouk, in accordance with the policy spelled out by General Prapas at a news conference in 1968 that the hillpeople in general should abandon their 'nomadic' existence.

By 1968, some 40 per cent of the upland population in Nan province had become homeless, while the number of refugees in the hills had risen to 15,000 by mid-1972 (Thompson 1968 a, b, c; Thaxton 1971; Cooper 1979; Tapp 1979).

It was at this time that the government most clearly revealed its fears of insurgency among the non-Thai minorities and a policy of assimilation towards them. It is not the intention here to provide an exhaustive account of the emergence and problems of these programmes over the past thirty years.[12] Since this book is primarily concerned with the *assimilation* of the White Hmong and the definitions of Hmong ethnicity which are employed to aid or act against such assimilation, I have rehearsed these details here in order to show, first, the kind of pressures and external agencies which have bearing on the lives of the Hmong today, and, secondly, to demonstrate the extent to which Thai government policy towards the Hmong and other uplands-dwelling minorities in Northern Thailand has been in principle assimilationist.

Moreover, there is no doubt that such attempts to destroy the inaccessibility, or autonomy, of upland villages, have exacerbated rather than relieved ethnic tensions in the area. Pressures on land since the early 1970s have resulted in sporadic outbreaks of violence between Hmong and Karen settlers. However, as one responsible informant told me, 'We had no problems with the Karen before the *kolong* (Northern Thai)[13] came here. The *kolong* built roads to the villages, and with the roads came the *kolong*. And the *kolong* brought the problems' (Nomya, October 1981).

Many Karen interviewed, on their side, showed a fundamental admiration and respect for the Hmong and the culture with which they were frequently in conflict.

The Impact of Educational Programmes

Both in the main settlement of Hapo and in the focal village of Nomya, primary schools had been established under an innovative and potentially creative scheme of 'non-formal education' which received some international assistance. Despite a profusion of ideas and an observed commitment to local problems on the part of some of the teachers, who were in the main members of other minority groups themselves, although never of the same groups they taught, these schools evidenced many of the problems I have described earlier as endemic. The villagers had initially opposed the idea of a school, seeing it as a way for outsiders to get a foothold in the village. It was said that the school abstracted children from the fields when their parents needed their help at critical harvest times, and distracted them from the play and games which had been their traditional prerogatives. Although officially attendance was not compulsory, in effect it was so through coercion. And although one of the merits of non-formal schooling in theory was precisely that attendance at school should *not* be demanded when economic imperatives demanded otherwise, in practice attendance was only not expected when the teachers were absent, which did not necessarily coincide with village priorities. One of the most important aims of this type of schooling was to instil in Hmong children a sense of Thai nationalism and responsibility towards the state. For example, the words 'We are Thai people. We must learn the Thai language', chalked in Thai on a blackboard, were chanted and copied regularly by all the children. Although basic geographical, historical, and arithmetical education was imparted, the main value of the school was to act as a centre for the dissemination (and collection) of official information. The teachers functioned generally as spokespersons for the Thai authorities, recording births and deaths, conveying official pronouncements, and occasionally convening village meetings. In structure, the schools consisted of open, thatched wooden buildings with seats, a blackboard, and a large Thai flag hoisted above. On one occasion, I watched a child shamed to tears because of his muddy appearance by being made to stand on a stool beside a child who wore slightly cleaner clothing. It is very difficult for Hmong children who work in the fields to avoid being a little muddy! Although instruction was by no means brutal or unkind, it could as in this instance become a vehicle for prejudice and ethnic antagonism, and such schools are an apt symbol of the way local projects tend to function in minority areas.

Such projects are, moreover, an example of the way ethnic minorities have been treated by their lowland, majority 'hosts' over a great number of years. The underlying aim has been to integrate and assimilate, if not destroy, the culture which differs from the majority of the state, and to deprive it of its economic base. Consider the following account of a field-trip to a remote village in the mountains of north-west Guangdong in the 1930s (Fortune 1939):

Since the founding of the Republic, China has been spending some money towards the assimilation of the Yao. The Bureau for Civilizing the Yao builds secular school houses in the Yao villages, takes a few Yao children away from the mountains, educates them in the Chinese secular schools, and returns them to the mountains as future school teachers. The Yao prefer for the present to decline the facilities offered their children by these school teachers and school houses. They prefer their children to work in the fields, and for some of them to acquire a non-secular education offered by Yao priests, in recitals of the names of the gods, and of books of incomprehensible nonsense, which, in the mouths of the priests, is believed to cure the sick![14] There is money made by the priests, and an opening for the successful pupil as a priest. The priestly 'education' is compatible with children's work in the fields, however, and is not realised by the Yao to be nonsense. It is their own, by previous adoption, and they prefer it. Secular education would take more time, and an opening for a successful pupil in the Yao hills is not viable.

We know that similar work continued with the Miao, the people most closely related linguistically to the Yao, and similar to them in many elements of their culture and economy (and had, indeed, preceded the Republican era).[15] The basic means whereby the Thai government and its agencies have sought to implement its strategies towards its upland-dwelling minorities, however innovative they may seem and however effective they may be in their initial impact upon particular local populations, in fact represent highly conservative strategies which have characterized relations between state populations and upland-dwelling minorities of the region over a period of several centuries. It is in this that we may seek some of the reasons for the retention of the village's sense of autonomy and conceptual inaccessibility noted above.

Lombard-Salmon (1972) has given an extremely interesting account of the policies of the Chinese government during the eighteenth century in the province of Guizhou, when farmers, miners, traders, and officials of Han Chinese stock were deliberately brought in to colonize and populate the remote mountainous areas which had formerly been the preserve of the Miao, under wide-reaching land-reform campaigns which, for example, forbade the Miao languages, proscribed the wearing of traditional Miao clothes among the men, and forbade them to cut their hair (in the process introducing the Manchu queue which can still be seen in Hmong villages today), and established schools to sinicize the young Miao (Lombard-Salmon 1972, pp. 220–9).

Obviously it would be unjust to suggest that recent Thai government projects are simply a continuation of centuries of persecution of the Hmong by their stronger, more powerful neighbours. Yet it must be pointed out that this is how many Hmong see it themselves, and that such views are enshrined in customary tales and legends of the past which we consider later in this book. It is true, moreover, to point out that assimilation has remained an abiding objective on the part of the stately majorities with regard to their upland-dwelling minorities. Assimilation has remained a constant concern of Thai government policies towards the Northern minorities, and this has affected government projects to do with reafforestation and the eradication of opium in the area today. It follows, that in this may lie some of the reasons for the many problems and failures which have plagued such projects. Were governmental development policies not so clearly linked with strategic policies of assimilationism, they might have had a far greater likelihood of success. And where assimilation is not in itself the major objective of a particular government project, all too often it becomes the manner in which it is administered, owing to the individual feelings of the extension workers who implement and oversee projects.

Thus it is through the intention to assimilate the minorities that the majorities have traditionally expressed their relationships with and attitudes towards the minorities, and similar policies of assimilationism pervade the rationale and implementation of governing majority policy towards the minorities today. Since, however, the minorities have so far *not* been assimilated, and indeed have strongly resisted such assimilation, none more so than the Hmong, it follows not only that there are internal contradictions within majority policy which act against the sort of assimilation it is designed to facilitate, but also that equally complex mechanisms have been evolved by the minorities themselves, in order to retain their political autonomy and ethnic identity. It is to the elucidation of these complex mechanisms of absorption and adaptation that this book addresses itself. For not only are there situations where Hmong villagers have fled across borders and into communist regions purely in order to be able to practise their customary modes of subsistence and worship. There are also situations, as in the village surveyed, where a very strong sense of autonomy— and, as we shall see, to a large extent a real autonomy—was maintained despite the very evident presence of Thai government projects which are described below. First, however, let us consider the account of another traveller (Kemp 1921):

Next day we reached Tatingfu, where we had already written to all the missionaries in charge to summon as many different tribespeople as they could for the weekend, and they assembled in good numbers from a radius of about forty miles. They have a flourishing school at Tatingfu, where the boys are taught Chinese as well as the ordinary elements of education, and about forty students varying in age from about ten to twenty had come along the winding path to meet us some two miles outside the town. They had gathered magnificent deep red rhododendron heads, which with their blue robes formed a gay picture. They all wear Chinese dress in the school, though they come from various tribes, as it is more convenient in the city, which is of course Chinese. . . . The city is surrounded by a wall, and is thoroughly Chinese. Formerly they had great trouble with the neighbouring tribespeople, but they say they have had much fewer disagreements since the missionaries have made friends with the tribes, and there seems to be no difficulty about their coming in considerable numbers to the mission premises, where they stay for the weekend. The majority of those we saw were the *Wooden Combs* (Ching Miao). . . .

Here we encounter the significant role which Christian missions have played in educational projects for the upland minorities and particularly for the Hmong, which we examine in more detail later. Kemp (a keen painter and botanist) is referring to the Green Hmong, and missionaries associated with the China Inland Mission, which began operations among the Chinese Miao at the turn of the century. When we turn to consider the importance attached to the acquisition of literacy by the Hmong, and the role which notions of its deprivation have played in formulating inter-ethnic distinctions, we shall be able to understand why it was that the very similar present-day missionary establishments described in Chapter 4, run by foreign missionaries of various denominations in Chiangmai, played a vital role in *reducing* the villages' inaccessibility from the external world, while at the same time preserving and increasing its immediate distance from the Thai state.

The Impact of Agricultural Programmes

Two agricultural agencies maintained a presence in, or near, the villages. One of these, a government (PWD) agency, had a station in Hapo, and loaned vegetable and fruit seeds, such as persimmons, apples, pears, and beans, as well as coffee seedlings, to the villagers, in an attempt to persuade them to abandon opium poppy cultivation, besides maintaining a model coffee garden which had gone badly to seed. Reaction towards the personnel associated with this project, which was in its fourth year (coffee having been planted for three) were particularly unfavourable. The villagers said they did nothing to help them, and complained of the cultural chauvinism of semi-officials who would only spend a few days in the village each month.

The second project, part of the agricultural programme associated with the King, maintained a post separate from and outside the villages, where it performed USDA-funded experiments with various alternative crops to poppy, in collaboration with the agricultural faculties of several Thai universities, in particular Kasetsart and Chiangmai. This project donated seeds, particularly of 'cool climate' crops, such as lettuce and broccoli, carrots and peaches, to villagers, and marketed the produce for them. Although reactions were more favourable towards this semi-voluntary project (which covered Nomya) than to the official government one (which covered Hapo), many complaints were forthcoming about tardiness in payment for produce received, exorbitant prices for fertilizer, and so on. A criticism often levelled at the 'cool climate' strategy is that, in general, the Thai do not eat such crops, and thus they only benefit the small foreign resident community and tourists. While this is true to some extent, the major problem was that most of the produce rotted on its long journey to the lowland markets. Despite these projects, which had been in existence for four to seven years, alternative crops had made as yet few inroads into the traditional economy, in which the production of opium as a cash crop continued to be integral.

Table 3, based on villagers' estimates, coupled with my own observations, and checked against Royal Project records, shows that the economy of the village remained a mixed one, and that there had been no serious challenge to the authority of the main crops in the annual cycle. According to the records of the PWD for 1980–1, however, which were not broken down by household, 10,500 coffee plants had been distributed among the forty-two listed households of Hapo (there were, in fact, forty-eight house-holds), and a further 2,000 in the focal village (twice as many as reported by villagers, since half had died). A further 420 mangoes were reported as planted by the combined villages, 505 lychee plants, and 320 peaches (ar-ranged through the Royal Project), while 5 rai had been planted with other crops which were not sold.

Even these figures, which may be an overestimate, represent insignifi-cant amounts of alternative crops when one takes the total number of villagers into account, and, moreover, neither leafy vegetables nor the type of cool climate crop favoured by the Royal Project, such as persimmons, had made serious inroads into the economy of the village, since many of them could be planted in opium fields, interspersed with maize, or in the dry rice swiddens. Binney (1971), for example, listed radish, parsley, cauliflower, six types of cabbage, mustard-grass, seven varieties of squash, potatoes, turnips, garlic, ginger, taro, six kinds of beans, peach, pomelo, banana, lychee, mango, and other fruits grown in such swiddens or else in garden plots near Hmong villages. Many of these, together with a variety of medicinal plants and those classed as 'good for feeding pigs', were evi-dent in the swiddens near the focal village, particularly radish, kale, dib (a kind of gourd), leafy cabbage, and tomato.

Only coffee and apple, if planted in maize/opium swiddens, would have had a serious effect on the cultivation of poppy, since in time they would take over the land, although the villagers were either unaware of this or did not recognize it as a danger, possibly because the maximum period for which such fields could be cultivated was often as few as three years, and this with some use of lime fertilizer on the barer and steeper slopes which reportedly resulted in higher and more abundant crops for a time, but in opium which was less good to smoke. Coffee, moreover, is an extremely difficult crop to grow and requires constant attendance and watching and

TABLE 3
Crops in Cultivation in Nomya*

Name of Villager	Dry Rice (rai)	Wet Rice (rai)	Corn (rai)	
Vaj Xeeb	4	2	10	130 apple trees; ½ rai lettuce; ½ rai cauliflower; some wheat.
Suav Yeeb	4	8	8	60 coffee plants; 124 apple seedlings; 50 pears; some persimmons.
Yis	6	4	10	60 apple trees; 50 coffee plants (¼ rai); ¼ rai lettuce.
Tsuj Ntxawg	10	–	12	50 coffee trees (for 3 years); 90 apple trees; ½ rai cauliflower; ¼ rai lettuce; some beans and wheat.
Nkaj Suav	12	–	14	—
Nkaj Huas	6	–	8	80 apple trees; ½ rai coffee trees.
Ntxhoo Pov	6	–	6	50 apple trees; ½ rai coffee trees.
Neeb	4	–	6	—
Nyiaj Paiv	2	–	12	—
Siv Yis	4	–	12	¼ rai lettuce; ¼ rai cauliflower.
Xeev Hwm	–	7	8	150 apple trees; ¼ rai lettuce; ¼ rai cauliflower; some wheat.
Ntxhoo Xab	10	–	10	40 coffee trees; 60 apple trees; ¼ rai broccoli, lettuce, and cauliflower.
Looj	6	–	8	¼ rai lettuce and cauliflower.
Tsav Xwm	6	–	8	¼ rai lettuce and cauliflower.
Vam Nrhuag	10	–	8	¼ rai lettuce and cauliflower; some beans.
Vam Pov	2	–	6	—
Nom Tuam	8	–	6	50 apple trees; 40 coffee trees; ¼ rai lettuce and cauliflower.
Yis	4	–	6	—
Caiv Phab	10	–	6	¼ rai lettuce and cauliflower.
Vaj Neeb	8	–	12	25 apples; ¼ rai lettuce and cauliflower; ¼ rai coffee; some persimmons.
Tooj	6	2	6	50 apples; ¼ rai cauliflower; some beans.
Pobzeb	6	–	4	—
Tsuj Xwm	4	–	2	—
Khav	6	–	2	—

Note: While officially covered by the Royal Project, Nomya villagers had also borrowed seed and stock from other villagers, as well as purchasing from the PWD at Hapo.
*At the beginning of the agricultural year 1982.

fencing. The history of the attempts to replace opium production with that of coffee in Thailand prior to 1985 was a sorry one and not essentially different from the earlier 'favourite' alternative crop, kidney beans. Cooper has documented how UNPDAC, in an attempt to guarantee a market for kidney beans introduced into tribal areas in 1970, had to buy up the entire crop at a cost of US$87,130, out of a total budget of US$2 million for a five-year period, an experiment it could not afford to repeat the next year, so that the fields reverted to opium (Cooper 1979). In examining the impact of coffee substitution projects on the UNPDAC 'key' Hmong village of Khun Wang, out of 15,350 trees given to the village between 1974 and 1977, only 1,445 trees were found to be left, of which only 956 were healthy (Lee 1981, p. 229). This may serve to give some idea of the problems faced by the villagers in my own survey with regard to the introduction of such crops into their annual cycle, which we examine in more detail later. Only four villagers reported making profits from the sale of coffee in 1980 in the focal village. Nkaj Huas had produced 458 kg of sellable coffee arabica, from which he made 6,141 baht, but this was exceptional. Ntxhoo Pov had produced 278 kg at 3,722 baht, Yis only 97 kg at 1,398 baht, and a member of Suav Yeeb's household had produced 93 kg at 1,102 baht. These were all members of the Vaj clan. Another two had experimented with planting coffee, but produced too little to sell.[16]

A number of points emerge from the data given in Table 2 which are of particular interest. First, it is obvious that serious attempts had been made by the villagers to comply with the wishes of the authorities. Over two-thirds had made some attempt to plant alternative crops. Planting apple trees, which would not mature for a decade and need guarding and weeding, to say nothing of coffee (which requires four years before yielding beans), crops for which marketing, transportation, yields, and prices were all fluctuating and uncertain, required a considerable effort from those villagers who did so, and lost valuable man-hours needed for actual subsistence. Secondly, all fields classed as 'corn producing' were those in which poppy was rotated in sequence with corn. Given the natural reluctance of farmers to divulge details of their yields and the very difficult bureaucratic situation with regard to poppy cultivation I have briefly outlined in the foregoing sections and expand later, it is only in certain situations that one may make direct enquiries and inspections of opium acreages, but corn fields here may be understood as a euphemism for opium fields. We see, then, that the great majority of villagers cultivated far more maize, in combination with poppy, than they did dry rice.

Three cultivated approximately equal amounts, while all five who culti-vated less 'corn' than rice were members of the Xyooj clan. This was not a matter of deliberate strategy, but rather reflected the very weak position of the Xyooj within the village relative to the dominant Yaj and Vaj groups, and their more restricted access to the lands about the village. The sample of those who cultivated less poppy than rice included the two poorest men in the village—Pobzeb, and the Khmu'.[17]

Both were in fairly wretched states, lacking goods and labour, and addicted to opium. It is very clear from the survey, then, that the village was not sufficient in rice, that it *was* heavily dependent on the sale of opium as a cash crop, and that the various projects donating and encouraging a variety of alternative crops had done little to mitigate this situation.

The Place of Rice in the Economy

We see, too, that a small number of villagers (five) cultivated irrigated rice fields, one to the extent that he had cleared no dry rice fields that year. However, he was old, and had many sons who did cultivate dry rice. The following is an extract from my field diary (Nomya, July 1982):

'Ua Liaj' (Cultivating Irrigated Rice Fields)

Of the new wet rice fields Vaj Xeeb's family are making this year at Poom Xami, only one field altogether is being done, of some five rai. After fifteen days' work we have only finished about one rai[18] and one nga, but the rest is fenced and will be left for another year. It was not forested, being where the cattle were previously. It is only planted after the first year, being terraced and steep; the first year only people (not buffaloes) work it, and it has been planted already and a stream diverted to it. Only Vaj Xeeb's family work it—they have no water-buffalo, only cattle, but these will do as well.

Of Xeev Hwm's wet rice fields which 'niam' (Vaj Xeeb's wife) went to help with yesterday, he has worked them for three years and they are quite flat. Originally he paid four vam (40,000 baht) to buy them from two Karen at Mae Sala. Apart from the old man's daughter, niam, her son-in-law Tub (Thoj) with his elder brother Pov and younger brother Tsuj Pos (Vam Neeb's son) and his wife, and Vab (Vaj Neeb Xyooj's son) helped on the principle of one day's work for me is one day's work for you. Tub has no wet rice fields, Vab only a little, and Tsuj Pov cleared one this year. This year also Vam Choj has rented wet rice fields—one—in which he has planted ten poom seeds. He has rented it for one year, at 2,000 baht, from a Karen at Huai Tong by the reservoir, where the fields are. It is very far, and he has nobody to help him, like Vaj Xeeb's fields at Poom Xami.

While the cultivation of irrigated rice strains does necessitate, therefore, a form of exchange labour which has been remarked for wet rice agriculture elsewhere,[19] the cultivation of irrigated rice land by those who can afford to do so does not represent a new strategy for the Hmong, as is often supposed, nor may its adoption in the focal village be seen as a tendency away from shifting cultivation. In Hmong economy, concentration on opium as a cash crop, or corn as a cash crop, or wet rice, depends very much and probably always has done, on the type of land available. The situation at Nomya was thus very different from one in Nan province, as the following account illustrates (Interview, Paklang resident, December 1981):

Twenty Hmong Lis families now cultivate only wet rice fields, which they need work only in the rainy season, but can grow vegetables and peas on when unplanted with rice. Only smokers cultivate poppy for their own consumption, and the settlement has become self-sufficient in rice. This arose out of a particular situation, since the land in that part of Nan was extremely dry so that it was too hard to work the fields and the residents had to go to another village to grow their maize. Nine years before, one family sold their crop of maize and beans for a profit of 20,000 baht, which enabled them to invest in some land suitable for irrigated rice cultivation (at that time the price of land was only 2,000–3,000 baht compared with prices of over 10,000 baht). A related family bought as much as 13 rai. There was considerable opposition, initially, to such investment, on the part of those who favoured the solution of relocating the village to a more fertile site. However, after the first year, many families came to help, at 15 baht per day per person (now 20 baht).

But this was an exceptional case, determined by the location of the village, and not the outcome of government projects to replace poppy culture with permanent agriculture. As a particular economic strategy, wet rice cultivation has clearly been an option since well before the inception of developmental efforts. Thus I interviewed many Hmong from the settlement sites who remembered their grandparents, and even in some cases their great-grandparents, cultivating irrigated rice.

Kiab, a woman who grew up in Sayaboury province of Laos, remembered her family cultivating wet rice there since the early 1950s on a small scale, while my assistant vouched that his mother's father had cultivated wet rice for many years before his mother had married his father, which takes one well into the 1940s. In Northern Thailand during the 1930s, there were still some old Hmong who remembered the use of the plough, and when asked why they had given it up, an old chief replied (Bernatzik 1947, 1970, p. 480):

This land, and even more the country that we had to pass through to reach our present dwelling sites, is mountainous, stony, and does not permit the use of the plow. In our old homeland, too, there were such regions in which one could not use the plow. In time the old men die, the young people no longer know the use of the plow, and so the custom begins to sink into oblivion, even where it could still be used today.[20]

This clearly attributes the choice of wet or dry field agriculture to the terrain. And it is also clear that wet rice cultivation formed a traditional option. Abadie (1924, p. 159) refers to increasing cultivation of irrigated rice in terraces by the Hmong of Tongkin. Lunet De La Jonquière (1906) also attested to its practice among the Hmong of Tongkin, while Betts (1899–90) had chronicled 'tier upon tier of paddy fields' built by the Miao of China 'driven into the uncertain fastnesses'. The situation could not have been put better than by the Hmong informant who told Savina, 'We Hmong, some of us only cultivate [dry] fields, some of us only cultivate wet rice, and some of us do both' (Savina 1930, p. 215, author's translation).

Irrigated rice cultivation is documented for the Hmong by Savina (1930, p. 216), Lemoine (1972a, p. 54), and Yangdao (1975, p. 76). ⌐ 24

It is important to make this point (also made by Cooper 1984, p. 34), since it would be all too easy to undertake the kind of analysis of Hmong economy which would focus on the disruption of a traditional subsistence economy and its exploitation by new market forces symbolized by the introduction of roads, or to class the rudimentary evidence of irrigated rice cultivation I have documented as demonstrating the inroads of new forces or even the result of alternative crop programmes. In fact, as we have seen, this is very far from being the case, since wet rice cultivation is rather a fundamentally conservative strategy adopted wherever resources are sufficient and available. The Hmong economy is, and always so far as I am aware has been, a mixed one, centred primarily neither on the production of opium as a cash crop nor on that of rice as a staple diet. Moreover, the lack of self-sufficiency in rice in the villages surveyed cannot be taken as signs of the exploitation of a traditional subsistence economy,[21] simply because, contrary to the theses of Geddes (1976) and Binney (1971), rice has always been something of a 'special food' for the Hmong.

At the very most, it has only been in certain locations and at certain periods that rice has formed either the staple diet or a major item of cultivation for the Hmong. Thus, in the south Chinese mountains we are told that, 'Rice is a luxury, as it will not grow on the high mountains, and the people can seldom afford to buy it' (Chinese Recorder L 1919, p. 803).

Hudspeth (1937, p. 99) follows S. Pollard (1919, p. 64) in listing the Miao diet at that time as consisting of steamed wheat, maize, oats, millet or buckwheat, and vegetables, with salt very scarce. Similarly, according to Graham (1954, p. 33), who worked in Sichuan during the 1930s, 'the principal food is corn meal cooked like porridge, but *those who are able to get it eat rice*' (author's emphasis).

Likewise, in Laos and Tongkin, rice was not initially a major food or crop for the Hmong. Bourotte (1943, p. 35) noted that the Hmong preference for maize *instead* of rice was 'exceptional' in Indo-China. Roux (1954) also noted the Hmong preference for maize in Laos. However, after the Second World War, Yangdao (1975, p. 72) has charted the changeover among the Hmong of Laos from eating maize to eating rice. Although his argument that the Hmong were 'once a people of the plains', and rice culture cannot be borne out solely on the basis of the linguistic evidence of the existence of a vocabulary in Hmong relating to wet rice agriculture, it does suggest that wet-rice agriculture had been known and practised for sometime (Yangdao 1975, p. 79). But clearly it can only have been in small quantities, since *rice was not the staple diet* for the Hmong in China, nor for those in Burma, where Scott (1900, p. 599) documented maize as the staple diet for the Miao, nor in Laos or Vietnam, as we have seen, until after 1945. In Thailand, however, rice seems to have formed the staple diet of the Hmong since at least the 1920s, and there is a rooted dislike for eating maize, which is seen as animal food.

The reasons for this become very clear in that, as Keen (1978) points out, the effective upper limit for dry rice cultivation coincides with the effective lower limit for poppy at 1 000 m. Corn, buckwheat, and other cereals all grow better than dry rice at higher altitudes, and when we compare the topographies of the different countries through which the Hmong have passed, we can chart the transition from maize as a staple diet to rice. For the mountains of Thailand are, in general, lower than those in the neighbouring countries, which range up to 2 800 m above sea-level in Laos, 3 200 m in northern Vietnam, and 7 600 m in southern China. Thus, the reason for the 'exceptional' preference of the Hmong of Indo-China for corn before the Second World War must have been in part the result of their relatively recent migration into the area, since only after 1945 was there a greater tendency for the Hmong of Laos to settle at lower altitudes.

It is doubtful, therefore, whether traditionally the Hmong economy was *ever* a self-sufficient economy centred on the production of rice. Rice seems not to have formed the staple diet in most of the previous habitations of the Hmong, and wherever it has, irrigated forms of rice cultivation appear to have been practised where possible, in addition to the shifting cultivation of dry rice. Thus, against the common developmental argument that it is rice deficiencies among the Hmong of Thailand which have brought about their greater engagement in the production of opium as a cash crop, I would argue that this cannot be historically accurate. Apart from very particular locations, there have probably always been serious deficiencies of rice production in the uplands and highlands. This has necessitated the concentration in the uplands on the production of goods which were scarce in the lowlands, and which could be exchanged for luxury items from the lowlands, such as the cultivation of tea by the Palaung of the Wa states of Burma (Leach 1954, p. 235).

Such exchanges contributed to instituted relations of dependence between the uplands and lowlands which were far stronger in the past than they are today, and which were disrupted by the advent of colonialism in the lowlands. The autonomy and isolation of the highlanders has been greatly exaggerated as a result. It is dubious whether *any* of the hill areas were 'really 100 per cent self-sufficient' in the past (Leach, personal communication, March 1984).

The Agrarian Calendar

According to the emic calendar, the first to second months of the lunar calendar are the time for slicing the opium. From the third to the fourth months, one prepares the fields and crushes the stubble and weeds, and burns off the forest. The fifth month is the month for planting the rice and corn (when everything begins to grow after a little rain). During the sixth month, one hoes the corn, and likewise in the seventh month, one hoes (and clears) the fields. The eighth month is the month for sowing the opium, and also for weeding. From the ninth to the tenth months, one should complete the (first) weeding of the opium. During the eleventh month, one should begin to reap the rice, while in the twelfth month, one finishes the reaping of the rice and celebrates the end of the year. The reaping of the corn is not mentioned here, because it is obvious that it must be done before or during the broadcasting of the opium seeds.

In what follows, I have not attempted to give an accurate or comprehensive ethnographic description of the labour process or the cultivation cycle. Instead, I have attempted a brief impressionistic survey of the whole sweep of the Hmong year, to give some idea of what it feels like, as well as of the major priorities and decisions confronting the villagers. In fact, the structural outlines of the calendar over the past forty years have remained remarkably traditional, and thus it would be easy to schematize a generalized, idealistic calendar for the entire Hmong economy. However, this would not convey the extent to which different households did different things at different times. According to their labour resources and priorities, or accidents such as illness or laziness, different households found themselves, for example, burning fields off after the first rains had fallen, or harvesting rice and late corn simultaneously, or even, as in the case of two individuals, tapping their opium poppies before the New Year celebrations because they had planted early and were in need of the produce.

Already, in June of 1981, the season of mango and durian, when flame of the forest and jasmine flourished in the lowlands, it was rainy and extremely hot. Dry rice was dibbled in the mountain fields. By the end of the month, it was difficult to travel to the mountain villages, except on foot or by motorcycle. No cars could go, and the roads were effectively closed. August and September were the hardest time of work in the year. The opium fields had already been cleared, and bunches of hemp were being stripped. Timber was collected, and cane sugar eaten. Tsav Pheej's lettuce fetched 2,700 baht in Chiangmai, and many people came to the village to buy opium. This was the most labour-intensive time of the year. Corn was being collected from the fields, and mushed up for sticky cakes in the large wok ordinarily used for pig food. It would not keep if sugar was added, so none was. There was not much to eat but red, broken ('Karen') rice and some boiled vegetables. Some rethatching of the roofs was carried on. One joi of opium fetched 4,000 baht. The children played, as the Thais do, with fighting dung-beetles on sticks of sugar-cane, and the broadcasting of opium began after the stones had been removed from the fields. Black soil was thought best, but the reddish soil around the village would do as well. Only one last, decaying corn field had not yet been plundered, and already (16th September) the greenish dry rice fields were in process of growth, with the felled trees still lying on the ground amid the plants (fixing the soil). One satang[22] of opium cost 40 baht. The refugees in the camp were short of opium, and much was sold in Chiangmai. It was a few days after 18th September that opium planting began in earnest, and was still continuing on the 28th. The women wove strips of hemp tucked into their belts as they walked around with babies on their backs. Vaj Xeeb collected wet rice. Pine was collected from the forest, but one tree would yield very little tinder. Sugar-cane and pumpkins, too, were collected. The opium fields, which were not yet planted, were beautifully turned after the plucking of the corn. Txhiaj Xauv roasted some potatoes. By the 7th October, corn was still being cut, the rice had not yet been harvested.

By the 17th, small leaves had appeared between the dried stems of the corn which had still not yet been cut—interspersed with radishes, kale, and gourds. Only a few families had completely finished cutting the corn before 25th October, but on the 30th the old man of the village offered to the ancestors the first new rice—well in advance of the proper harvest. Anything sold at this time was sold directly; it was only after the opium harvest that debts were repaid. During November, rumours spread that the UN and the BPP were coming to cut down the opium, and towards the end of the month the first weeding of the opium took place as rice was threshed by some in the fields. Five days of heavy rain marked the week up to the 25th November. By the 28th, some poppies were coming up in the fields—some bulbs, mostly green. The new long rice grains were perpetually winnowed in the houses, and cakes were made out of rice and wrapped in banana leaves for children and adults alike. From the 10th to the 14th December, the opium had to be thinned—it cannot be transplanted. Tooj (Xyooj) finished weeding. Opium was waist high in one field only, with glorious red and white bulbs. Towards the end of December, it became clear that the opium was not going to be cut in the area, 'because of the King', rumour said. The rice had mainly been harvested and threshed and collected home by pony, human back, and truck. Wood was piled outside the houses. Most of the opium was waist high. The New Year was celebrated, the 26th December being the first day of the new first month of the New Year. A water ritual on the morning of the first day told us that there would be little rain in the coming year. In the hot season after the cutting of the opium, people began to say, everyone will be free and there will be lots of weddings. By January 18th, some opium had been tapped, but most were guarding it against thieves. By the 20th, everyone was tapping opium. Opium cost 5,000 baht in Chiangmai—more than in the settlement, where it was 3,500 baht and elsewhere in the less remote mountains, where it was 4,000 baht only. The large settlement began tapping the opium several days before the focal village, having planted it a week before. But even at the focal village people had begun to burn the fields off, cooking the ground to improve it for planting, which would normally be done after the opium harvest. However, it took until March for all the opium harvest to be completed, and the fields began to be properly prepared, and by 3rd March it had actually rained a little. Vaj Xeeb collected opium from all his lineage brothers to give to his brother, a recent defector from the CPT, and shared out land he and they had already cleared with him and his family. Opium was weighed constantly in both the Chinese houses in the large settlement. By the end of the first week in March, all householders were stamping dry tapped opium pods for seeds for the forthcoming year. 'This season is the hot season, the Hmong are free and have no vegetables to eat with rice', people said. The fields we cut were mostly old rice fields not used for a decade, some two hours from the village. On the 8th April, after hearing a 'plas' bird call twice, the omen of the new year, for those who attended to such things, was partly reversed, since if this bird calls twice in the hot season, rains will be plentiful and all the crops will grow abundantly. The great winds which herald the advent of the first rains began to blow across the mountains as the valleys below grew hotter. Now the fields and scrub were being burned off, but those who had not done so in time were dismayed by the arrival of the first rains. On 8th April came a little rain, and some had not finished burning off. All the roofs had to be mended and the fields fenced wherever possible. In the mid-afternoons, frequent thunderstorms disturbed the bumpy red dusty paths. By mid-April, it had rained some more, some fencing was still in process, and the planting of the corn had already begun. Felled trees lay smoking in the dry rice fields. The men began to gamble in the evening. Since it was the hot season still, everybody had money. At last the real rains came—thunder and black skies. Some burning continued, while fields were still prepared for planting. The fields were fired in several places at once, with brands or matches, and then one had to run for it. A frenzy of fire-wood collecting took place, to beat the oncoming rains. It took less than a week on average for a family to finish planting their corn, and by the 25th April most had been planted. Some trips to other villages were made to look for and collect herbal medicine. More great thunderstorms occurred throughout May, and by the first week of June the torn rain clouds entirely blocked the peaks of the mountains, and the mud of rain became torrential rivers along the paths. Gourds were made into drinking utensils, and back-baskets woven together, and there was a shortage of water in the village. The roads were too bad for a severely ill woman to be taken to the medic in the neighbouring Chinese/mining settlement by her husband. By mid-June, Txos' corn fields were already knee high, and fields further away from the village were shoulder high. When weeding, one started upslope, and worked slowly downhill. Those who had wet rice fields began to harrow them, mustering what help they could. Towards the end of June (the 26th), both corn and rice fields had to be weeded, and some had started, very early, to strip the corn down. Exchange labour continued in the irrigated rice fields. By the 3rd July, new wet rice fields were being made and most of the corn simultaneously being cut. There was little to eat, but mushrooms, snails, and bamboo roots were gathered from the forest to accompany the rice. On the 4th July, the first man in the village offered to the ancestors the first corn cobs, although mostly only glutinous corn was being cut, and, at the same time, the rice fields needed weeding for the second time. By the end of July, the rains were easing a little. On 2nd August, the corn was being cut, and the fields were being cleared for opium again.

Outlining the major transitions of the year in this way leads to an undue emphasis on planting and harvesting which disguises the extent to which weeding is by far the most predominant economic activity. What I have tried to do here is to present a particularized picture of the cultivation cycle which must, therefore, necessarily be incomplete, not only because of the differences between individual households or because of my omissions of the many minor activities which make up the daily life of cultivation—such as collecting a back-basket of spinach from the swidden for the midday meal, or catapulting a bird and mending paths on the way to the fields in the morning—but also because it happens that the cultivation cycles of different hamlets themselves are to some extent deliberately staggered so that their New Year celebrations do not fall on the same day, and villagers may therefore attend the celebrations of another village after having performed their own ritual observances for the New Year. Thus the celebrations in the large settlements occurred several days before those of the focal village, while the refugees along the Lao border celebrated almost a month before that, 'because they have no rice to harvest', but also because of periodic adjustments of the lunar calendar (Lemoine 1972a, p. 50).

Apart from the anxieties associated with the unpredictable timing of the rains, which are vital not only for rice planting but also for opium (since too heavy rains occurring towards the end of the monsoon can wash the opium seedlings clean away), and the fear of thieves and predators (one villager in Hapo had his entire opium crop eaten up by goats), it will be seen that a major concern during the course of the year was whether the government would come and destroy the poppy plantations, on which so much time and work had been spent,[23] or whether the villagers would be allowed to reap their own harvests. I describe this below. A further point to note is that a comparison of the emic calendar and my own survey above with more formal calendars assembled by other researchers (E. Heimbach 1979; Binney 1971; Lee 1981; Yangdao 1975) show that the structural outlines are broadly similar. This is remarkable, given the extensive influences of twenty years of development programmes I have documented earlier, and the influences of other external agencies, and exemplifies the manner in which the village had managed to retain its own conceptual structure, its own sense of its own autonomy, as one might expect a calendar to do through its function of relating a conceptual cycle to the actual cycle of events (Evans-Pritchard 1940, p. 100). And this was despite the very real inroads of what cannot be described simply as a market economy, since it seems that the Hmong have been involved with a market economy for a very long time,[24] but at least the forces and influences associated with the developing capitalist Thai state.

Slash-and-burn Policy

Since what follows is a negative case-study, that is, a case-study of something which did not actually happen, one will have to remain aware of the importance of rumour in Hmong communities which rely primarily for their information on oral means of communication (cf. Firth 1967). These included, in particular, Hmong friends in Chiangmai who were in contact with branches of the government and aid agencies, local broadcasts in Hmong from the radio station in Chiangmai, and the project officials, agricultural and educational, stationed in or near the settlements.

It was not until October 1981, a month after the opium seeds had been broadcast in the fields, while the first laborious weeding of the tiny shoots was taking place, that rumours began to circulate in the village and in the settlements, that the government was coming to 'bomb' all the opium plots that year. It was said that the UN and the BPP were to be the implementing agencies, and at first that all opium fields everywhere were to be fire bombed or otherwise destroyed indiscriminately. I was asked if this was true, since there was little the villagers could do about it now that they had planted the poppy, and the project officials had let it be known that the rumours were substantially correct, but had given no details of when or where the attacks were to take place. In fact, it was true that by the end of October, the Office of Narcotics Control Board (ONCB) had exchanged 41 kg of poppy seedlings for bean and sweet corn seedlings in some fifty hamlets which had undergone intensive development activities over a number of years. Moreover, a plan had been announced by the Substitute Crops Division that police would be sent to ten more villages to destroy the poppy fields during November. It was not yet clear, however, what extent of involvement there would be by the military, defence volunteers, or village scouts at that time.[25]

A situation of extreme unease was created in the villages, which increased throughout November as rumours of which ten villages had been singled out, leaked out. The officials of the King's and agricultural projects again confirmed the plan, but refused to confirm which the ten villages would be. This created further frustration, fear, and anxiety about opium production among the villages falling under their jurisdiction. One of the village teachers, however, took the part of the villagers, arguing that destroying opium would be all right if the villagers had anything else to grow, but that they did not, only wheat and apples.

In the latter quarter of November, an incident occurred in a nearby village which resulted in the death of a Thai police officer and the jailing of one Hmong. It was said that the police (to whom bribes are sometimes paid before major transportations of opium are made) had informed the Hmong of one village that they would buy some opium from them at a particular site in another village, but had then stopped the truck conveying the opium at a point midway between the two villages and attempted to commandeer the shipment without paying for it. The Hmong had surrounded their car in order to prevent this, and in the ensuing shooting one police officer was killed. It seems clear that this incident resulted from the great unease caused among the Hmong by such rumours, and also that the arrangements under which the local police coexisted with opium producers had been placed under serious strain by the official government policies favouring the immediate confiscation or destruction of opium crops.

On 27th November, it was announced that eighty 'volunteers' from Chiangmai were to be paid 100 baht a day to destroy selected poppy sites under police scrutiny, while helicopters were to be arranged to visit the more remote areas of poppy cultivation. At the same time, rumours began to circulate in the survey site regarding me, to the effect that I was in the employ of the Thai government and had been sent to the village to conduct a survey of the poppy fields to be destroyed, hence my predilection for visiting poppy fields. These rumours, traced eventually to a Chinese shopkeeper living near the mining community, were fortunately systematically contradicted by my assistant, who made a point of ridiculing them and assuring the villagers that, on the contrary, I was interested primarily in shamanism and geomancy.

Most of the villagers, however, did not believe these rumours in any case, and I was able to discuss with a good number of them how they felt about the rumours that the poppies were to be destroyed, which by this time had been lent a lot of official sanction. The headman of one of the villages was, perhaps, the most forthcoming and articulate about his reactions. The Hmong, he said, were well aware of the evil effects caused by opium elsewhere, thanks to the instruction of local officials and the Chiangmai radio station, and had made earnest and repeated efforts over the years to plant anything which they had been encouraged to plant which could provide them with the income at present generated by opium. One crop after another had been tested, he said, with no success, and it was his opinion that the local projects were not really serious about wishing to eradicate opium production, since so little real help had been received from them and it was so obvious that there simply was no effective alternative to the cultivation of poppies. If the government *had* been serious in its wishes to destroy the opium that year, he asked, why had they not informed the villagers *before* they had planted their poppies? And he quoted a certain Catholic priest he remembered visiting the village telling him that the 'Hmong were the first people for suffering in this world'. This headman had been elected to his position primarily because he was a liberal, tolerant, and gentle man with a great interest in the outside world and new developments. Moreover, he maintained more extensive contacts with his Karen neighbours and the local structures of the Thai bureaucracy than many of the other villagers.

Other villagers, however, lacking these liaisons, were less comprehending, more fearful, and more explicit in their reactions. The most generally stated feeling was that they would have to defend their crops by force if they were threatened. With very little attachment to land as a commodity, villagers nonetheless saw their *crops* as their own personal property, which in any case had normally to be constantly defended against thieves and predators. However officially sanctioned the destruction of poppy fields might be, in Hmong eyes such actions would be illegitimate and unjustifiable, and the only defence available, as it is on other occasions, was violence or the threat of violence. Another rumour then began circulating, this time to the effect that I had been sent by the English government to 'help the Hmong' in various ill-defined ways. This, too, had to be denied, as I repeated that I was there for my education. The threats of violent reprisals by Hmong villagers, however, were by no means ill-founded at this time. It is well known that some heroin refinery owners supply hillpeople with sophisticated armoury in exchange for their opium produce. It was even said that the local communists would help if nobody else did and, if approached tactfully, might be able to supply the villages with anti-aircraft guns, while the roads could easily be blocked and made impassable with felled trees. This idea may have started because at that time nine Hmong defectors from the CPT were due back to the region, and two had recently approached the villagers in the settlements to ask if they might settle there. I was asked by the villagers to write letters to Bangkok explaining the situation, which I did. However, in December it was announced that the plans to destroy most of that year's opium crop had been cancelled owing to certain 'faults' in the programme, and because it was feared that violence might ensue. In the form in which rumours of the announcement reached the village, it was said that there had been a personal conflict between the Premier, who had favoured the plan, and the King, who had intervened and vetoed the plan on humanitarian grounds. The rumours reached the village before the official announcements, however, and seemed to allay much of the apprehension, since poppy fields had been cut in designated areas in previous years with forewarning, and this, it now seemed, was all that might be done. The fact that there was official recognition of the likelihood of violent resistance showed good judgment of the situation on the ground. However, the position of the establishment was redeemed to some extent in local eyes through the attribution of the cancellation of the project to the personal intervention of the King.

It transpired, however, that there had been a disagreement of opinion between the ONCB, responsible for supervision of all the programmes for opium eradication, and the Royal Project, which had maintained stations for many years in many of the areas where poppy was cultivated, and which favoured a more gradualist approach. Although opposed from the first in Bangkok because it was felt that the moves would 'antagonize' the hill-dwelling minorities, 140 rai in Chiangmai had still been scheduled for destruction by over 1,000 BPP and defence volunteers during November. The programme was delayed owing to strong opposition. Finally, in the first week of January, it was reported that seven villages voluntarily destroyed some 100 rai of their own poppy fields, while a further three villages had refused. Some of the slashing was personally inspected by a high-ranking team from Bangkok, including the Deputy Premier and the Secretary-General of the ONCB. Pictures were distributed to both the local and international press.

There is no doubt that serious international and particularly American pressure had been put on the Thai government to reduce that year's bumper crop of opium (estimated at 50–100 tons).[26] The week before the poppies were destroyed, the US Assistant Secretary of State for International Narcotic Affairs had visited Bangkok, and later in January the Thai Ambassador to the United States, in Thailand for a one-month consultation session, was due to visit the sites where poppy had been destroyed. As the Hmong in the villages had correctly averred, the UN office had favoured the plan, and after it had been shelved, general accusation was directed at the inadequacies of the UN crop substitution programmes in having failed to ensure, after a decade, that its key villages would have sufficient alternative crops to maintain subsistence levels if the poppies were destroyed.

Although the official announcement was that the villagers in the areas of destruction had voluntarily destroyed their own crops, and had been given substitute crops and rice with which to support themselves, the interpretation given by Hmong in the survey site was rather different from this. According to those at least in the focal village, the very few Hmong who had actually cut down their own poppies (calculating 10 rai to the average opium producer would mean that the 100 rai destroyed represented the fields of about ten producers) had been well compensated. The event was, therefore, seen as something of a public relations exercise, effected after it had become apparent that the original plan was unworkable owing to the probability of violent resistance by villages which thus retained a very real autonomy, despite their participation in the Thai state.

After the cropping, at a special meeting of the Premier with national security officers, the ONCB, the Foreign Minister, and representatives from the Departments of Agriculture, Forestry, and Irrigation, it was decided that a greatly expanded programme of substitute crops would have to be effected in addition to the already existing UN project. In a press interview, the Secretary-General of the National Security Council (which is responsible for counter-insurgency operations in Thailand) condemned what he called the 'slash-and-burn' approach towards opium eradication, and suggested the establishment of co-operatives to loan money, provide rice, and market cash crops on behalf of hillpeople to be deprived of their poppy crops in the future, which would have to be grown on land made available by the Forestry Department.[27] However, the Forestry Department, committed to a radical policy of reafforestation, has, in general, been reluctant to make land available for alternative crop programmes.[28] Here one may glimpse some of the extremely complex issues involved by the above case.

All parties involved in this case have particular concerns and interests which often run counter to those of other parties, and the accommodation between these different interests is so delicate and so unstable that any radical innovation may lead to severe repercussions. The victims, of course, remain the Hmong and other local-level producers, since they are largely at the mercy of these interests. In retrospect, it seems clear, first, that growing international pressure was largely responsible for the introduction of the original plan, and, secondly, that the various sectors of the Thai administration displayed a characteristic tact and finesse in the various stages through which the plan was put, since the demands of external aid agencies had to be carefully balanced against a local situation in which heavy-handed intervention could do nothing but harm. The dissemination of rumours of the proposed destruction of poppy fields well in advance of official announcements certainly ensured that an atmosphere of alarm and tension was created in which it became feasible to pull out of a plan which was at that time impracticable. As the headman had complained, if it had really been intended to implement the plan in the first place, it would have been far easier to disseminate the rumours *before* the poppy seeds had been planted.

The suggestions which finally emanated from a number of different authorities, to the effect that the proposed 'co-operatives' would require a considerable amount of international financial support in order to become viable, demonstrate the vulnerability of the opium grower in terms of national issues. Pecuniary considerations remain primary at every sector of the system involving opium production—from the local poppy cultivator concerned with ensuring that his family has enough to eat in the coming year, and the district police official maximizing his personal benefit through playing off the demands of his duties against an existing situation of extensive poppy cultivation, to the senior officials who manage and control developmental and strategic policy in the region. In the final analysis, one may note that even here, where financial considerations play such a crucial role, it was ultimately strategic considerations which determined developmental policies. The plan for enforced eradication of large areas of opium poppy was not primarily called off because of the political opposition there undoubtedly was. Primarily, the plan was called off because, as the Hmong revolt of 1967 had showed only too clearly, and as was clear from the inception of the plan, such policies were not practicably enforceable in the hills, and would almost certainly have led to widespread resistance if pursued. Thus the village *retains its autonomy*, at the centre of a complex web of bureaucratic intrigue and vested interest. As one Hmong put it to me, of the events of 1967: 'The government bombed us because the government was stupid. It is no use bombing small people like the Hmong.'[29]

The following paradoxical interview with the Governor of Chiangmai, who gained considerable popularity at the time through opposing the plan, illustrates very clearly the complex dialectic between dependence and autonomy, which operates at an international level between Thailand and its trading partners, as it does at the level of the local village (*Bangkok Post*, 20 December 1981):

Foreign support for crop replacement has been very weak. . . . It's easy to destroy the poppy plants . . . but the Government is concerned over the problem of how the hill tribes are going to live when their poppy plants are destroyed. . . . If we say that we are driving the hill tribes into the arms of the Communists, we will be accused of being alarmist. . . . We cannot eliminate opium planting 100 per cent. When we move against the hill tribes in one area, they go to another. . . . If we don't set up administrators we won't be able to stop them from their nomadic wanderings to destroy more forests. . . . Even in areas which have been developed the hill tribes continue to grow opium. They don't do it in the villages but at a distance from the villages. . . . The American suppression officers don't understand. . . . Look at the opium wars of the past and the interests of foreign countries which the opium trade served. . . . The way to attack the problem at root is to assist us in crop replacement. . . . They set up tariff barriers against our commodities . . . it adversely affects our national economy. . . . If these and other countries want us to destroy our poppy plants and continue these kinds of trade policies, they'd better forget it. . . .

One has to remember that this interview came after, rather than before, many years of crop-replacement schemes.

Alternative Crops

In one UNPDAC Project village I visited, where crop-replacement efforts had been far more intensive than in the villages under survey, I was told that there was no rice to eat, and it was impossible to plant rice or corn because the site was so exposed that the seeds were blown away. This also meant that houses needed rethatching twice a year, a time-consuming if not expensive business. Of alternative crops they had planted, red beans had failed, and it had been impossible to sell peaches because of lack of transportation. Nor were the villages permitted to grow opium. 'Whatever is planted has to be replaced by something else,' added this informant.

Gar Yia Lee, an anthropologist who is also Hmong, studied the impact of UN development policies on the socio-economy of another Hmong village, also a key UNPDAC site, and provides the following typical complaint by one of the villagers there (Lee 1981, p. 223):

I went to the project office in Chiangmai to ask one of their representatives to come and supervise the harvest as we were told in the beginning. The representative said he had to go to another project village but would come to buy the potatoes from me later. I waited seven days but he did not come, so I and my family dug up the potatoes and took them home, all very big and without diseases. I then approached the extension workers at Khun Wang here, but they said the potato price was very low so I would have to wait until the price improved and they would then let me know. But I never heard from them, and after a few more approaches, I stopped.

In the end, some of the potatoes were consumed by us, some by the pigs, but most just rotted. The following year, we grew again, but could not sell any, even at the local market, so we just gave up on potatoes.

It should be noted that these were crops which the UN Project had persuaded villagers to grow on the express understanding that the Project would buy back the product at guaranteed rates.

Let us inspect the case of Tsav Pheej, one of the residents of the larger settlement, and the villager who had taken official exhortations to concentrate on the cultivation of crops other than opium poppy the most seriously and consistently. Tsav Pheej was an imaginative and outspoken man, and thus unsuitable for the sort of position of power occupied by the headman, which required a more discreet and diplomatic personality to liaise with the Thai authorities. However, his voice carried considerable weight in the village, since his status was central to the axis of power there, which ran from him and his younger brother to his two paternal first cousins, one of whom was the current headman. He was also among the most sociable of the villagers, a great gossip, always visiting other people's houses and advising on problems which did not concern him directly. He was respected and looked up to in the village as well as further afield, and spent more time talking to the women and playing with children than most of the men. In his youth he had been a strong adherent of the communist party, fighting for many years against the government before surrendering and returning to live peaceably in the village. He spoke good Northern Thai, Shan, Chinese, Lahu, and Karen. One of his sons was in prison for transporting opium, and another was studying in Chiangmai. Tsav Pheej complained to me (Hapo, October 1981):

For many years I trusted the government and worked to show that I was a good man. I planted lettuce, beans, tomatoes, and coffee in over 6 good rai which could have been used to plant opium. That was three years ago, when the price of opium was 1,200 baht. Every month in the rainy season I look after a hundred kinds of vegetables and fruits. But it is impossible to sell them. I have taken them to Bo Kaeo, Chiangmai, and even the Karen villages, but nobody wants to buy them. Only the Kaset (agricultural project) will buy, but at very low prices. There's nothing one can do about it; one has to wait a long time before they pay up, sometimes many months. Now I think it is better to grow opium, but the land is no good.

He was fortunate in having been able to acquire an extent of land for his vegetables which was relatively flat. While it was said that level land could be used for opium for up to twenty years, it was rare for a cultivator to possess as much as a single rai of this, while steep land could be planted at most for three years with dry rice, and then three years with maize in sequence with poppy, but not with good yields. For example, some old fields cleared by a household from the focal village, which had not been used for ten years by the Karen who had originally cleared them, produced very small rice plants indeed after three years of cultivation. The Hmong seem always to have planted a variety of vegetables in their swiddens, and maintain plantations of fruit trees around their villages. The economy is, as I have said, a mixed one (cf. Lemoine 1972a), which is not primarily centred on the production of opium as a cash crop, but includes important elements of animal husbandry (Table 4), hunting, gathering, and the cultivation of other plants, and requires a viable cash crop to maintain levels of subsistence. Thus, the attempt to introduce vegetables and other crops into the economy could not have represented a serious challenge to a major crop unless there was a regular and guaranteed market for it. The essays of men like Tsav Pheej, together with the fact that opium cannot have become a cash crop for the Hmong much before the turn of the century, while the maize which constitutes such an integral part of the economy today was a New World crop only introduced in the sixteenth century, show that here, contrary to the assumptions of development economists, we are not dealing with a traditionalist subsistence economy at all, hedged about with all sorts of superstitious taboos. The Hmong have not been slow to innovate in the past, provided that the innovation was reasonable.

The Hmong have a considerable, if grudging, respect for those whose cultures are seen as more stable, more centralized, and more literate[30] than their own, and they must pay heed to the exhortations of Thai officials and

TABLE 4
Livestock Ownership by Household

Household	Chickens	Pigs	Cattle	Horses
2	49	13	20	13
3	60	14	40	8
4	55	16	2	3
5	26	9	2	4
6	27	10	1	2
7	29	18	8	4
8	20	4	–	1
9	60	12	36	4
10	15	8	6	4
11	18	6	–	2
12	21	7	3	6
13	30	5	–	–
14	90	13	–	4
15	15	12	3	3
16	28	12	2	7
17	12	5	–	1
18	37	13	45	10
19	16	8	–	4
20	6	3	–	1
21	16	7	8	2
22	14	6	–	–
23	15	4	1	1
24	27	6	–	–
25	8	3	–	–
Total	694	214	177	84

foreign experts to concentrate on the cultivation of these alternative 'cash crops'. Yet it has become very clear that these crops, such as beans or potatoes, many of which the Hmong have always grown, cannot be called 'cash crops' if there is no market for them. Neither do the arbitrary markets, set up by development agencies at artificially standardized rates work well, without the sort of massive external subsidization which has in the past been lacking.

The Opium Trade

Given these sorts of 'risks' and 'uncertainty', it is difficult to see how a producer in such a situation could possibly calculate even 'expected' utility with any degree of accuracy (Heath 1976). Certainly, it would be inaccurate to depict the Hmong as maximizing their returns from a position of perfect knowledge of and control over available alternatives. Opium has become socially embedded in a way that alternative crops have not. Moreover, demand for opium must increase indefinitely owing to the addictive nature of the commodity. Hmong opium production results largely from the demands of an external market and trading mechanisms. McCoy (1972, p. 135) has documented how the influx of Guomindang (KMT) army remnants into Thailand in 1961, following the Chinese Revolution of 1949, who began to monopolize opium transportation, contributed to the massive increase in Thailand's opium production. Guomindang remnants, together with other para-military organizations which in the past played a useful role in suppressing communist insurgency, are still settled along Thailand's northern borders, and with their associates retain a primary role in the opium trade.

Profits from the Thai Royal Opium Monopoly, which imported opium into Thailand and was only brought to an end in 1958, averaged some US$5,500,000 as late as 1950 (Ingram 1971; Table 5).

P. Cohen has documented how the authorization of poppy cultivation in 1946 in areas of Northern Thailand led local Shan, and later Chinese and Northern Thai traders who already had a long history of trading with the hillpeople, to 'exploit the burgeoning commerce in opium' (P. Cohen 1981, p. 88). One of the five types of opium trading isolated by Cohen is described as follows (P. Cohen 1981, pp. 86–91):

August and September was the peak period for the exchange of rice for opium. Swidden rice was not harvested until November, so by this time Hmong rice stocks were either depleted or in short supply.[31] Also the Hmong needed extra rice to feed Karen workers whom they hired to hoe fields for the planting of opium. A number of Baan Talaad traders bought milled rice at Chiangmai to exchange for Hmong opium. At the Hmong villages of Mae Win (the upland sub-district of Sanpatong) the exchange rate was forty *thang* of milled rice per *joi* of opium. However, by August or September many Hmong had exhausted their opium stocks either by sale earlier in the year or by smoking. They were therefore forced to buy rice from traders by means of a credit arrangement in which they agreed to hand over opium to the trader after the harvest between December and February. For such transactions the Hmong received only twenty *thang* of milled rice for one *joi*

TABLE 5
Government Revenue Sources

Source	Year										
	1892	1905	1906	1915	1917	1926	1927	1938	1941	1944	1950
Direct Taxes											
Land Tax	1.02	3.86	7.58	8.22	8.68	11.62	12.83	7.42			
Capitation	0.45	4.14	4.93	7.69	8.34	10.06	10.05	7.47			
Income and Salary Tax								1.85	3.30	22.87	111.98
Indirect Taxes											
Import Duties	1.74	2.69	3.01	3.49	4.18	7.16	16.03	30.58	33.76	11.75	557.75
Export Duties		2.79	2.86	3.67	3.70	4.05	5.00	5.46	11.17	3.38	170.56
Excise Taxes	2.30	4.16	3.95	6.22	8.62	10.87	12.15	6.98	19.68	57.70	356.62
Inland Transit	0.28	1.56	1.67	1.89	2.09	2.10	0.88				
Gambling Farms	4.28	8.64	6.65	7.46	0.33						
State Domains											
Forest Revenue	0.15	2.05	1.47	1.61	2.65	3.82	4.55	3.78	4.05	5.54	27.30
Mining Revenue	0.53	1.21	1.54	2.43	4.94	4.34	4.01	5.27	11.13	4.37	65.36
State Enterprises											
Opium Regié	2.48	10.26	8.87	16.56	21.18	18.01	18.18	10.39	14.27	61.33	110.15
Commercial Services		2.34	2.98	4.84	5.18	13.64	15.42	21.11	33.83	63.18	395.55
Rice Sales											140.00
Fees, Fines, Licences	0.26	3.54	3.89	5.56	7.60	7.17	8.20	12.64	16.68	20.01	118.09
Other	1.89	3.22	6.11	4.72	4.97	7.75	10.14	5.01	13.13	35.88	65.93
Total	15.38	50.46	55.51	74.36	82.46	100.59	117.44	117.96	161.00	286.01	2,119.29

Source: Ingram (1971).

of opium. This system of credit or mortgaging was called 'selling opium green' (*khaii fin khiaw*).

And so it still was in the survey district, together with the other system of 'selling rice green', whereby traders exchanged goods, usually with Karen, for repayment in *rice* (after the rice harvest) which could then be exchanged for Hmong opium.

But the Hmong in areas such as my own were fortunate. Not only did individual pedlars of various ethnic extraction travel up to the village at all times of the year to buy large quantities of opium from different house-holds, but two Chinese shop-houses were maintained in the larger settle-ment on a permanent basis, under the sponsorship of a third peripatetic Chinese who maintained his own house there, as well as several others (some with attached wives) in Burma, Chiangmai, and other areas of the hills. Not only did the nearby presence of mining communities ensure the persistent emergence of young risk-taking entrepreneurs who would make sporadic purchases of opium, but Chiangmai itself was close enough for most months of the year for villagers to be able to transport their produce there themselves. Moreover, there were no armed convoys of para-military organizations which visited the village on a regular basis demanding their share of the opium harvest, as there are in many border areas. The situation could not have been better expressed than by the villagers themselves, who pointed out that it was not opium which went looking for money, but rather money which came looking for opium!

When one has witnessed money come begging for opium, day in, day out, such a statement becomes impossible to interpret cynically, in support of a view of the Hmong as economic maximizers.[32] It is, instead, a simple statement of fact, which reinforces the view of the Hmong themselves, in their capacity as opium producers, as essentially at the mercy of market demands.

The major trader settled in Hapo was a Muslim Chinese who had orig-inally emigrated from Burma. He sold many kinds of minor goods—plastic sandals, torches, aspirin, dyed wool—as well as specializing in opium broking, for he was heavily addicted to opium himself. While no accurate records were kept, he estimated that he made 4,000–5,000 baht per month from his shop, and contrasted his present wealth, with one son studying Arabic at a Muslim school in Chiangmai and another helping in the shop, with the previous poverty which had forced him to take up residence in the village. His wife detested the village, complaining of the headaches caused by the vapours from rainfall on the animal faeces around the village. She described herself as a Khonmuang, but in fact her Khonmuang father had converted to Islam after marrying her mother, daughter of an Indian Muslim from Burma. This man's situation contrasted markedly with that of the other Chinese in the village, who did not make much profit from opium, for avowedly moral reasons but, in fact, largely because his solitary and self-sufficient personality had made him extremely unpopular in the village in relation to the first Chinese. As an addict, the latter would wel-come the men of the village to smoke, drink, and play cards in his shop after dark. Not being Muslim, but having left the Guomindang (KMT) militia in which he had been a regular soldier, the former also lacked the extensive contacts which the other had outside the village, and declared he was unable to make 50 baht a day.[33] He had lived by himself for many years until deciding to take a young Karen girl of fourteen (he was over forty-five) in marriage for company, and to help with the shop. The girl, however, who had been more or less sold into marriage by her mother's new husband, who was addicted to opium and only a few years older than she was, was miserable at the marriage, which was celebrated by the major Chinese traders in the local market town in the village. She ran away twice only to be sent back again by her mother and step-father, and eventually took to smoking opium with the addicted Khonmuang couple who lived on the outskirts of the village. While her husband was away on a trading visit, she succeeded in bartering away most of his stock in exchange for opium, and when he came back he was destitute. The girl left to work on a road construction project far from the village, and the Chinese, who had attempted to build a fish pond which had failed because, when the rain came, all the water ran out of it (since the bottom had not been lined) started to clear some dry rice fields near the village for subsistence. He was, in any case, heavily indebted to the peripatetic Chinese trader I have mentioned, through whose hands most of the opium produced in the district passed at one time or another, and who, together with the Japanese manager of the mining company, formed the effective authorities in the region. His son and the Japanese, for example, had contributed 20,000 baht to a road which was built to Chiangmai. Such Chinese referred to themselves in Thai as *punyai* (big people) in opposition to the Hmong, who were *punoy* (small people), and represented the widespread trading community which the Hmong farmers supplied.[34]

Until recently, the ban on opium production was not enforced for the humanitarian reason that poppy cultivators would starve if they were not allowed to produce opium for sale. This, however, encouraged graft and corruption and resulted in an extremely vulnerable position for them.[35] The practice of levying unofficial local tolls on opium production became widespread. The following extract clearly paints the picture (Report prepared for the BPP, 1965):[36]

The Meo have a history of unpleasantness with Thai authorities. Since they are uncitizened outlaws, living on land which is not theirs, violating forestry and opium regulations, the Meo easily fall victim of what they understand as injustice. Nearly all the *kamnans* in Chiangmai province extract a head tax of twenty *baht* or a small pig worth thirty *baht*.[37] The provincial police and occasionally the BPP will seize unregistered weapons of the hill people, yet never bother the armed irregulars.[38] Occasionally a family's opium production for the entire year-worth about 2,000 *baht*, will be confiscated.

This was the source of some of the anxiety I have tried to describe at the time of my research. Another reason for not enforcing the ban was because it was feared that alienating the Hmong and other opium producers would strengthen the widespread insurgency in the North. Since the recent collapse of the CPT's armed rural struggle, however, serious attempts have been made to enforce the ban by destroying large areas of poppy. While not rendering life any easier for the hillpeople, such a policy at least reduces the ambiguities inherent in tolerating a technically illegal mode of production.

Many development planners have adopted the argument that opium, which tolerates soils of lower fertility than dry rice and can thus be cultivated for longer periods of time on the same land, has been increasingly cultivated owing to the expanding population of the hills, which has brought about such a severe strain on resources that dry rice agriculture has become no larger possible (Geddes 1976, p. 133).

Not only does such an argument importantly underestimate the extent to which the hill-dwelling minorities have never been completely autonomous, self-sufficient practitioners of dry rice agriculture, but it also ignores the extent to which social forces connected with external markets determine the allocation of scarce resources and dominate the type and extent of production. The agent of appropriation is a human, not a natural, one. Quasi-ecological arguments that it is population pressure, rather than trading mechanisms, which is primary in the expansion of opium production, tend to support measures to reduce the population of the hills by the introduction of resettlement programmes or contraceptive projects.

As in Shanghai and Hong Kong (McCoy 1972), the illegalization of opium in Thailand has inevitably encouraged the production of heroin, which can be refined from opium relatively cheaply given the necessary chemicals, is odourless, ten times less the bulk and weight of opium, and thus more undetectable, easier to store and transport, and infinitely more valuable.

Reports such as the following (*Bangkok Post*, 4 May 1982) have become more frequent:

GUARD KILLED IN RAID ON DRUGS FACTORY

One Hmong hilltribe guard was killed and another was arrested during a pre-dawn raid on a mobile heroin refinery located in a valley in Samerng District. A Border Patrol Police force seized nearly thirty kilogrammes of morphine from the refinery, after a fierce gun battle with some twenty-five Hmongs guarding it. The clash at the refinery, allegedly owned by a Chinese Haw, came shortly before dawn yesterday.[39]

These reports do not convey, however, the extent to which villagers *themselves* are becoming addicted to heroin, or the emergence of an extremely grave heroin problem within Thailand itself.[40]

The failure of the proposals submitted to the American Congress in 1975 by the leaders of the Shan States in Burma to sell directly to the United States the entire opium production of the Shan States, showed quite clearly the errors of trying to suppress opium production 'at source',[41] since the real source should be sought in the social problems of the advanced Western countries which cause the market for heroin. We see that for the Hmong, as for other opium producers, an exceedingly delicate and complex relationship has been required between them and Thai officials who are of a different *ethnic* extraction from themselves, which has allowed the Hmong village to retain its conceptual autonomy at the expense of economic dependency. In considering this, we shall be led towards an examination of the very foundations of Hmong ethnicity itself, and the sort of cultural symbols they employ in order to express their differences from the members of 'other' cultures.

Developmental Views

Developmental projects have targeted the Hmong, in particular, as the major producers of opium. Yet the amount of opium produced either by the Hmong or by Northern Thailand is relatively insignificant in world market or even 'Golden Triangle' terms (since the majority of the Triangle's production emanates from outside Thailand, although it may be transported through Thailand).[42] The Hmong are not the only ethnic minority members in Northern Thailand producing opium, while, as we have seen, a large number of upland swiddeners are now ethnic Thai, who tend to practise the most destructive forms of swidden, neglecting, for example, to take precautions against forest fires while burning off their newly cleared fields, or not leaving the stumps of newly felled trees in such fields as the Hmong do, which helps to prevent erosion. Thus many of the effects of these projects tend to be interpreted by the Hmong as evidence of a fundamental ethnic hostility and antagonism towards them.

Since it is my contention that ethnic concerns have remained fundamental in development projects which have sought to transform the traditional economy of the Hmong, based on shifting cultivation, into a permanent-field type economy, we should here examine more closely some of the many charges made against this traditional economy.[43] First, it should be pointed out that shifting cultivation can contribute positively to the natural ecology. Indeed, the Hmong practise a geomantic system for the siting of villages and ancestral graves which demonstrates the fundamental harmony of man and nature (described more fully in Chapter 8), while it has been argued that shifting cultivation, which encourages a diversified economy, preserves and maintains the existing eco-system far better than a mono-culture such as one based on rice (Geertz 1963). Nor is shifting cultivation the main cause of deforestation in Northern Thailand.[44] Household fuel gathering throughout Northern Thailand must account for a very large proportion of this, since whole trees tend to be cut down for small amounts of timber. Large-scale logging and mining concessions operating in many parts of Northern Thailand are also responsible for deforestation.[45] Mining concessions near the survey site had, for example, made five good upland valleys uninhabitable and certainly uncultivable.[46] Pilot and test 'development' projects also take up a good deal of fertile (forested) land.

From the villagers' point of view, the role of proliferating development projects, mining and logging concessions, and reafforestation reserves, resembled nothing so much as a bitter contest for fertile land, a gigantic game of Go, in which the amount of land available for them to cultivate was constantly diminished. Swidden areas left to fallow were continually replanted with pines if left untended even for short periods, while teak was frequently planted on fertile land rather than at the elevations where it would have been of most use in fixing the soil. As early as 1973, this 'haphazard' and 'illogical' forest activity in the hills was criticized by a development expert (Keen 1973). As one villager put it (Hapo, August 1982):

The suffering and problems of the old days have continued up to the present regarding the planting of trees and bamboos. Neither Hmong nor Karen have land to subsist from. Although we have many children, we cannot go and cut down the forest, because this forest is only for planting beans in. The Forestry Department have a supervisory post near here (a post to watch the ridges), and they told us that wherever there was land good enough to cultivate, they would let us use it, and where it was no good for cultivating, they would give us trees to plant on it so that it would become a good place to cultivate. But now they don't do that at all. They just take trees and plant them everywhere.

When shortages of the ordinary water supply to the larger settlement occurred during the monsoon rains, which turned the entire earthen surface of the village into muddy torrents of water, the villagers were well able to explain to me that this was because the forest immediately above the village had been cut, leaving little but scrub to hold back the downpouring waters.[47] However, they added, most significantly, that this was *because* government restrictions forbade both the cutting of forest elsewhere, and the relocation of the village to a more forested site. In the past, they said, the Hmong had always taken care to leave a belt of forest surrounding their villages, and had always for preference cut well beneath the level of their villages. Although the Hmong in the survey site did not in general cultivate land on forestry reserve areas (although most upland is technically forest reserve land), their areas of plantation were severely limited by those areas where intensive pine planting projects were in process, and thus their fields were moving further and further away from the villages. Here again, as with regard to the cultivation of poppy, at any moment compensation may be demanded by forestry officials for swiddens which had been unwittingly cleared in forestry reserve areas, or informal arrangements may be entered into which allow swiddeners to cultivate such land. The fact that less than 20 per cent of the ethnic minority population of Chiangmai province possessed Thai citizenship, and the majority, as shifting cultivators, had no legal title to the lands they cultivate, increased the vulnerability of the minority shifting cultivator immeasurably in this respect. Thus, it becomes still more clear that the shortage of land in the uplands is by no means the sort of inevitable natural process, resulting from overpopulation, it is all too often depicted as. Shortage of land is the direct result of social divisions and ethnic inequities which favour certain types of occupancy at the expense of others. When we come to consider some of the myths and legends collected in the village, we shall see how the entire history of the Hmong is phrased in terms of similar mechanisms of territorial control, and inter-ethnic rivalry for sovereignty over the land.

Emic Reasons for the Cultivation of Poppy

Despite extreme fluctuations in the price of opium (Table 2), and the vulnerable position of the poppy cultivator with regard to central authority, opium still provides more economic certainty than any of the alternatives which it has been attempted to introduce into Hmong villages. But even apart from the insatiable demands of the market, and the willingness of traders to trek to hill villages to purchase opium, while more perishable foodstuffs and livestock must often be sold at less than half their real values owing to spoilage during transport (whole basket loads of chickens can die on the way from Nomya to Chiangmai), there are yet other reasons why the Hmong, if left to their own devices, would prefer to concentrate on the cultivation of the opium poppy. Not only is opium easy to store for long periods of time, as many commentators have remarked (Geddes 1976) and not only does the need exist to complement insufficient rice yields through the sale of opium, which has functioned for so long as a means of exchange in the highlands. There are also highly significant emic reasons for the continued cultivation of poppy, which remain to a large extent socially embedded in the constitution of Hmong society. Some of the reasons which I was given are the following.

The Hmong value silver as a sign of wealth, for personal adornment, and for the bridewealth paid at weddings, which is said to compensate the parents of the bride for the 'tears and labour' expended in bringing her up. Traditionally, the only commodity which the Hmong can exchange for silver has been opium, and as one informant put it to me, 'Without opium we cannot get silver. Without silver we cannot get wives!'[48]

Another reason frequently given for the continued cultivation of poppy was that because it is usually planted in sequence with maize, the disruption of the maize and poppy cycle demanded by surrendering poppy cultivation would result in a lack of the fodder for household livestock which maize generally constitutes—so that livestock would be drastically reduced.

Above all, however, opium still remains the most important medicine available in the Hmong village. I show later how health and literacy constitute the primary objectives of the Hmong and how village life is consequently dominated by the presence of illness and 'aliteracy' (cf. p. 124).

Opium as a pain-killer is effective against back pains, head and stomach aches, rheumatism, lumbago, and toothache. The effects of opium in relieving menstrual pains, as well as those associated with childbirth, are incomparable. Opium is also effective in curing the common symptoms of dysentery and gastro-enteritic disorders, which can be associated with cholerary parasites or other infections, and may be fatal, especially for young children (who sometimes have opium blown over them under a blanket if they are seriously ill). The anti-tussive effects of opium also stop coughs and sore throats, and can thus alleviate the symptoms of tuberculosis. Opium relieves mental distress, and can be smoked on social occasions, functioning as a 'social' drug like alcohol (see Westermeyer 1982). It is also commonly agreed that opium keeps one warm when the weather is very cold, although I can find no medical evidence in support of this. The high rates of addiction (nine of the household heads in the focal village) constituted another pressing reason for the Hmong to continue cultivating the poppy on some scale. A purple variety was said to be the best medically, and most cultivators maintained small patches of this primarily for their own consumption. The case of Xauv must stand as a typical account of the causes of opium addiction.

Xauv was a remarkable old man of fifty, extremely garrulous, hard of hearing, and considered slightly 'cracked' by some of the other villagers. When a Lua' passing through the village dropped senseless, Xauv was the only man in the village with the temerity to approach the man whom everybody feared was dead. (In fact he proved to be epileptic,[49] but people had feared to approach in case of *dab*, or the spirits, a sensible prohibition to observe in times of epidemic.) Xauv made his own rifles, to shoot 'tigers' as he claimed, although all I ever saw him shoot was a large wild 'leopard' cat. He lived alone with his wife, who was unusually affectionate towards him, in one of the smallest and poorest houses in the village, little more than a bare hut. He had been an opium addict for over twenty years, and his habit was so expensive (his wife was also addicted to opium) that he barely grew enough to support himself and his wife, and his entire life was spent in just scraping by, catching the occasional field mouse, squirrel, or bird. He was ill with a variety of complaints—chronic bronchitis and rheumatic fever, and a badly infected leg which made it difficult for him to walk. He had originally been given opium to alleviate some of the pain caused by a broken leg, and had become addicted to it in a matter of weeks. Once, twelve years previously, he had taken himself to Chiangmai and voluntarily put himself through the 'cold turkey' treatment for opiate addiction offered by a local hospital. On his return to the village, he had again fallen ill, with a high fever, and his wife, who also smoked, had offered him a little opium to relieve the fever. He had found himself addicted again. His present wife was his fifth. His first wife had been much older than him and, being told by a lineage relative that the child she bore was not his own, Xauv had sent her away. His second wife had left him because he had been unable to stop smoking opium. A third wife had died, and the fourth had left for the same reason as the second. He had never made any bridewealth payments. He had been poor all his life, and now his health was deteriorating rapidly. Once he badly needed medical attention for a tumour on his foot, and begged me to help send him to a hospital in Chiangmai. When everything had been arranged he backed out, saying he could not spend a night in hospital, because they would not let him smoke opium there. He was unable to confront this possibility after his previous experience in hospital. He used to suffer from terrible nightmares, such as being pushed into a deep hole by a tall being, and was convinced that he would die shortly.

It is in this sort of context that the choice of the average cultivator to grow, or not to grow, the poppy, should be viewed.

Notions of Authority

Although authority is ultimately invested in the past, and in a past which is seen as distinctively 'Hmong', there is no failure to recognize the *power* of the Thai state and its servants on the part of the Hmong. Indeed, Hmong attitudes towards officials of all kinds, are fundamentally tinged with admiration, envy, and respect, as the following extract from an interview (Nomya, August 1982) may demonstrate:

Freedom means doing what one wants. We would like to be able to do what we want, but there's very little of it. For the Hmong in Thailand there's nothing but suffering. None of our children have studied for a high degree, and there are still very few of us who have studied the laws. Some people would like their children to go and study but they can't afford it, because they haven't any money. I wish there was one Hmong to be the first among us, and then some of us could become officials, like that District Officer. We can only go on being slaves like this. If only there was someone who could study up to PhD level, but there's nobody who can [even] get into the government. To deal with illness you need to be able to study medicine, but if you have no money, how can you study?

I think that if we remained honest (of a straight heart), and had a supreme person to lead us, like General Vaj Pov in Laos, then the Hmong would become renowned, and things would be better than before. One would be content even if one had enough rice to eat, but now there are so many problems, that whatever one does is impossible, like if one cannot grow opium one cannot eat. I don't know what to do.

These Thai people are not like the Westerners. Westerners like the Hmong and try to help them, but the Thai will not listen to them. As for freedom, nothing can be done in Thailand except under compulsion, under the yoke of oppression, because whatever you do they tell you not to.

Those Hmong kings which have arisen occasionally were all murdered, and now there are no kings arising under heaven for the people of this world. We have to have learning beforehand, we must study books so that there will be wisdom and cleverness among us, so that everybody will see that it's appropriate for us to become officials, but slowly, slowly, it should be done, Hmong should help other Hmong and send their children to study highly and then Hmong will be able to help themselves.

The association between ideas of literacy, bureaucracy, and royalty, becomes increasingly clear in subsequent chapters. Here I am trying to make the point that, despite their resentment and grudges towards Thai officials, for reasons which have been considered throughout this chapter, the Hmong nevertheless demonstrate a respect and admiration towards them, and wish similar opportunities for themselves.

I well remember my classificatory father, Vaj Xeeb, telling me proudly that his son was going to 'be a policeman', when in fact he was attending a six-day BPP indoctrination course for teenagers in the local market town. The boy from Hapo who was studying in Chiangmai told me once that he was 'ashamed' to be Hmong when he saw the progress and development of the Thai city: 'Hmong live far from progress, with no land for their trees,' as the informant above added.

Suffice it to say here that the results of a questionnaire distributed among all the Hmong resident in the town of Chiangmai showed that only one was employed on a permanent basis by the UNDP, and he had run into such serious difficulties that it seemed he would shortly be forced to quit, while another was due to graduate from Chiangmai University with no serious prospect of employment before him (there were two others at the University). The latter had a brilliant mind, and had done so well at two stages of schooling, after having been sponsored by an American missionary, that he had continued into further study. Now the only alternatives for him were either business, probably employed by a Chinese trader, or to proceed with still further studies, which he could not afford. There is a considerable lack of vocational opportunity for the members of ethnic minorities in Thailand. The great majority in this survey were studying in Chiangmai, either at the Catholic school (with a few at Protestant establishments), or at Buddhist temples, with some at ordinary Thai secondary schools. Official employment was virtually non-existent.

Although there is no doubt, then, that the average Hmong maintains strong feelings of respect towards officials of every kind who are associated with the organization of states, at the same time there is a felt opposition, and distinction, between such state employees and 'ordinary people', such as the Hmong. In the fieldwork area, it was very apparent that the primary sympathies of the Hmong towards members of other ethnic groups were directed towards the Karen, the poorer Chinese or Shan, and those Khonmuang who were in a similar position with regard to centrally appointed Thai officials, and might also maintain dry rice or maize and poppy swiddens. In this sense, the real, felt opposition between local people and officials appointed with government sanction to guide and direct their lives overrode ethnic considerations. Nor was this only so for the Hmong. There was a recognized bondage of community between people who had to work the same kind of land in order to live, as opposed to those who did not, but received their salaries from some 'higher' authority. Karen and Chinese interviewed, whatever ethnic prejudices they also demonstrated, invariably showed a certain sympathy and respect for the Hmong as other people making a living, which was notably absent when local officials were discussed. Although they had to 'get along' with such officials, they were often discussed in tones of cynicism, and I think this is important in revealing the extent of opposition between 'the state' and 'the people' which still persists today, and certainly orders traditional inter-ethnic perceptions.

In truth, the clash over issues of deforestation and poppy cultivation does, unfortunately, express a more fundamental conflict between two very different kinds of social organization. I have talked with Thai agricultural officials who showed not only incomprehension but fear at the thought of a people so unwedded to specific fields as the Hmong, and with clan relatives scattered, as they saw it, across hundreds of miles of Thai territory. This sort of social system, based on shifting agriculture, contrasted strongly with that of the Thai, whose irrigated rice-fields might be passed down from generation to generation, and whose descent groups are far from ramificatory. Nothing better, on the other hand, expressed fundamental Hmong notions of freedom and egality than the indignation displayed when a proposal to introduce electricity and housing rent in a

UN project village near Chiangmai was mooted among villagers of the survey site. The introduction of electricity was opposed, not only because it was felt to be 'not really Hmong' and could prove dangerous during the rains, but in a very sophisticated way because it was felt that it would prove expensive, and thus lead to conflict or rivalry—*sib tw*—between those who could afford it and those who could not. (*Sib tw* was defined as one person wanting to be like another, for example, wanting a radio, too, in contrast to *sib txeeb*, which was applied to two people wanting and struggling for the same thing, which neither necessarily possessed.) The introduction of electricity would, it was felt, not be 'fair' or 'having enough for everybody'. As for the proposals to tax land at 5 baht per cultivated rai and 8 baht for a house plot, I was asked, 'How can they expect us to pay for the land? It is not theirs. It was there before any of us, like the air we breathe.'

1. I follow Heath (1976) in the analysis of situations of uncertainty, as opposed to those of risk or certainty, in decision-making processes.

2. 1 joi = 1.6 kg; 20 baht = approximately US$1.

3. In 1980, this was 12,365 baht (*Thailand 1982 Plan, Problems and Prospects*, National Identity Board, Bangkok 1982).

4. In 1965, the United States, through USIS and USAID's Office of Public Health, was responsible for retraining and re-equipping 1,600 Provincial Police and 6,300 BPP (extracts from a speech by Marshal Green, *SEATO Record* IV, 3 June 1965).

5. This became the Internal Security Operations Command (ISOC) after the normalization of relations with China in 1976.

6. For this and the following sections I have drawn on a number of sources. Full references will be found in Tapp (1979).

7. 'Communist Insurgency in Thailand: An Unofficial Summary of a Government White Paper prepared by CSOC', *SEATO Spectrum* 1, IV-2, 1973–4.

8. Prapas Charasuthien, 'Thailand's Hill Tribes', PWD Bangkok 1966.

9. See Geddes 1967; Mandorff 1967b; Davenport *et al.* 1971; Geddes 1983.

10. This entailed the construction of narrow roads between hill settlements and widening of existing ones: a notoriously difficult task regarding shifting settlements.

11. *Statement of the Policy on the Hill People's Development and Welfare*, PWD Bangkok 1976.

12. Tapp (1979; 1986).

13. The derivation of this word was traced for me by an informant in the village from Chinese *guo lo* (originally used to refer to the Lolo landlords of the Hmong in areas of southern China).

14. Fortune (1939) notes this religion as Daoist.

15. Betts (1899–1900).

16. One coffee plant is estimated to yield 1.0–1.5 kg of beans per year.

17. Poverty was calculated etically in terms of yields, availability of labour, and the possession of draught and domestic animals, and emically in terms of working for oneself, having others work for one, or having to work for others.

18. 1 rai = two-fifths of an acre; 1 nga = a quarter of a rai.

19. Moerman (1968b).

20. This quotation is also of interest for illustrating the 'theory of deteriorating knowledge' among the Hmong which we examine later (Chapter 6).

21. Indeed, the enforced sedentarization imposed upon communities of shifting cultivators itself has a great deal to answer for in terms of soil deterioration and the destruction of natural resources.

22. A weight approximating to a small coin now a hundredth of a baht.

23. Feingold (1975) estimates 387 labour hours are needed to produce just 1 joi of opium.

24. See Ross (1979) for the distinction between a 'communal' group characterized by traditional isolation and little awareness of its status with regard to other groups, and a 'minority' group, largely defined by its host population and with consequent lack of control over its own resources.

25. *The Nation*, 31 October 1981. External information in this section is primarily based on a series of news reports by radio and in the local press, later checked against interviews with senior narcotics officials attached to the UN in Bangkok and Chiangmai, and to a foreign embassy in Bangkok.

26. Mathijsen (1982).

27. *The Nation*, 13 January 1982.

28. In 1973, for example, 300 Hmong and Thai villagers demonstrated against a 3,000 rai forestry reserve project which had suddenly made them outlaws on their own land in the Mae Chaem area. This was during Thailand's brief period of parliamentary democracy (1873-6). Also cited by Cooper (1984).

29. Thus the weakening of the CPT, in the years since fieldwork was undertaken, has resulted in a very different situation as regards opium policy. The government can now, and has, taken serious steps against both the KMT and the SUA.

30. See Chapter 6.

31. Rice deficiencies were thus prevalent in the hills well before the 1950s.

32. See Godelier (1973) on *Homo Oeconomicus*.

33. He sold oil for lamps at 10 baht per bottle until he was undercut by a Hmong villager who started selling it from his own house after dark at 9 baht a bottle.

34. See the ECAFE- and UNESCO-aided *Report on the Socio-Economic Survey of the Hill Tribes in Northern Thailand*, PWD mimeo., 1962 (published September 1966) on the position of the KMT.

35. For example, with regard to mainland Chinese who had been repatriated from Thailand to Taiwan.

36. 'The Hill Tribe Programme of the Border Patrol Police', mimeo. Thailand Information Center, Bangkok, doc. no. 09186 (1965).

37. *Pujaibahn* is the village headman, and *kamnan* the sub-district officer.

38. 'Armed irregulars' here means the KMT militia.

39. The term *Haw* refers to the Yunnanese Chinese, usually Muslims. Its derivation is uncertain, but it is the same term as is sometimes transcribed *Ho* in Vietnam.

40. Bangkok Metropolitan Spokesman, TIC News, March 1983. I have spent a great deal of time assessing different reports of the numbers of heroin addicts and drug imports in Britain and the US and it is an impossible task. Official figures can only be based on *seizures* of heroin, and *registration* of addicts, while sampling-type research is usually too small-scale to be generalized. The only certainty is that official figures are invariably a gross underestimate, never an overestimate.

41. *Proposal to Control Opium from the Golden Triangle and Terminate the Shan Opium Trade*, Hearings before the Sub-Committee on Future Foreign Policy Research and Development, House of Representatives, 94th Congress, April 1975. Cf. Feingold (1975).

42. Comprising the opium-producing areas of Burma, Thailand, and Laos.

43. Just to give an example of how such identifications, against all reason, continue to infect others even than development planners, a recent anthropologist comments that the Northern Thai hills are 'populated by exotic tribal peoples who slash-and-burn and grow the opium poppy'! (Potter 1976, p. 12).

44. *Report of the United Nations Survey Team on the Economic and Social Needs of the Opium-Producing Areas in Thailand*, prepared by J. F. O. Phillips, W. R. Geddes and R. J. Merrill (Government House Printing Office, Bangkok, 1967).

45. Cooper (1979, p. 16) notes the forestry industry as the primary culprit in Northern Thai deforestation. Illegal log-felling alone accounts for 30 per cent of deforestation by official estimates.

46. Together with mining, road construction was the major cause of erosion in the hills, argues a recent NADC Report. 'Poor road construction is probably the major cause of soil erosion in the watersheds,' according to the HASD Inception Report (1981). These are recent views, however.

47. Lee (1981) cites further research as showing that most run-off caused by deforestation occurs just when it is most needed for padi, early in the rainy season, while the later run-off which causes flooding tends to come from the forested areas.

48. Elements of 'ceremonial exchange' (Herskovits 1940, p. 447) are also involved quite apart from commercial transaction, since opium can be used in gratitude for shamanic services, offered to guests, or in part payment of bridewealth obligations.

49. *Lombamu* in Thai: for some reason it was not described in Hmong.

3
The Political Dilemma

THE political alternatives for the Hmong, during the period of fieldwork, and as seen by them, may be summarized as loyalty to the King, or support for the CPT. We have already seen how, to some extent, these alternatives present themselves in the course of the fundamental economic alternatives of whether or not to produce opium. It was shown how government policies towards the eradication of opium, because they are so clearly connected with assimilationist concerns, have led directly towards resistance to the state. This did, in fact, happen in the case of the ethnic rebellion of 1967, and might well have happened again in the uncertain context of the 1981 poppy cropping episode. Here we consider to what extent instrumental and deferential concerns play a part in the allegiances which have been formed in certain areas between the Hmong and the CPT. First, however, it will be useful to examine the other side of the coin: support for the King, as the symbol of the Thai state.

The Thai State

Recent years have seen a conscious attempt by the King and his advisers to reinstate something of the traditional relationship which formerly obtained between lowland monarchs and those upland-dwelling minorities who accepted their jurisdiction. In the traditional Tai States, certain hill-dwelling ethnic minorities were revered as 'original holders of the land' by the valley-dwelling wet rice cultivating peoples who had established political dominion over them. Their acceptance of the dominion of the rulers of these states was symbolized by offering annual tribute to them in the form of forest products, such as ivory and teak. In some areas, ritual games or battles were established in which minorities would be annually defeated by representatives of the state. Although I can find no evidence of similar relations in the Chinese Empire, the Jarai minority played a similar part in the Khmer Empire, and it is possible that the Tai rulers adopted such customs from the Khmer. In the past, such ceremonial relations of dominance and dependency between ethnic minorities and lowland states were typified by the Lua' minority of the Tai state of Chiangmai, the Wa minority of the Tai state of Chiangtung (in modern Burma), and the Khmu' of the Tai state of Luang Prabang (in modern Laos) (Kraisri 1965; Archaimbault 1964; Mangrai 1981; Scott and Hardiman 1900). A special position was thus assured these minorities at the periphery of the state, which depended on their relationship with the sovereign of the state, under whose direct protection they nominally fell.

However, the formation of modern nation-states in the region, through the British colonization of Burma and the French colonization of Indo-China, severed and disrupted the traditional dependency between states and the inhabitants of the remote regions, which was based on the reciprocal exchange of goods. The Shan states were incorporated into the nation of Burma, just as Luang Prabang and the other Lao states were formed into present-day Laos and North Vietnam. Chiangmai, after the appointment of a Siamese Commissioner to its Court in 1874, was formally incorporated into the Bangkok-based state of Siam (which first became known as Thailand in 1939) just after the turn of the century. As the power of traditional lowland monarchs yielded to that of a centralized modern nation-state, inevitably, the conceptual distance between ethnic minorities and majority populations was increased and widened. The King's much popularized sympathies towards the mountain people, evidenced in the agricultural schemes launched under the 'Royal Project' in 1970, and assistance given to BPP schools before then, and furthered by the Queen's patronage of hilltribe handicrafts, represent an attempt to reconstitute this damaged relationship between the traditional Tai monarch and ethnic minorities.

By contrast with most other development programmes, the personal style of patronage adopted by the Thai royalty has been surprisingly successful. It is fair to say that, like most Thai until very recently, Hmong in general do respect the Thai King, and reserve their disapprobation in general for his officials and bureaucrats. Geddes (1976, p. 106) has described the large summer palace maintained by the King on one of the mountains near Chiangmai, quite close to two much-developed Hmong villages. The King makes a point every year, accompanied by a royal entourage and press photographers, of visiting selected minority areas in the hills and talking personally to the villagers. Posters of the King are common in Hmong villages, in particular one showing him in conversation with a Hmong headman on an equal basis—the only time such a picture may be seen in Thailand, where people meeting the King are expected to keep their heads respectfully lowered.

One household in the larger settlement was inhabited by an extremely impoverished couple with a daughter who had been badly crippled from birth. The Hmong look after such children, and care for them as for others. She could crawl around the house on her hands and feet but suffered from an extremely bad speech defect (she could, however, embroider). Both parents were opium addicts, and the man would hire himself out to other villagers during house construction or rethatching for a few baht per day, which was not a common practice. Their daughter had been presented to the King on one of his annual visits to the region, and had been sent by him for several weeks for treatment to one of the best hospitals in Bangkok. This treatment had been ineffective, however, and the parents did not seem particularly to appreciate it, but rather accepted it as part of their general lot. Token though such gestures may be, there is no doubt that they do have a recognized impact on Hmong conceptualizations of the Thai state and their place in it, which is not insignificant. A story circulated in Thailand about a Khmu' tribesman who, finding the King had visited his village in his absence, decided to return the call, and was admitted to the palace.

The King employs a number of minority artisans on a regular contractual basis at his palace at Phuping, and the second son of a household in the focal village was gratified to have been promised a contract there as a silversmith, during the course of fieldwork. The headman, however, remembered an incident which had taken place two years previously when he had been presented to the Queen in company with the supervisor of the local Royal Project station. The Queen had asked him the price of a kilo of beans in Chiangmai. Before he had been able to reply, the project official had answered that it was 10 baht, although it had in fact been only 5 baht. The headman blamed the official for this lie, and associated it in general with other false information about the prices of alternative crops, but was nevertheless demonstrably proud to have been visited by royalty.

Thus, the political loyalties of the Hmong towards the Thai state have in the past been largely expressed through the medium of a *personal* conception of the sovereign of that state. The significance of this is brought out in the following examination of Hmong notions of royalty. There is a strong tendency to envy the Thai for being powerful enough to have a king, and likewise to attribute the social and economic disadvantages of the Hmong within Thailand to their lack of a strong leader with central authority, and the divisiveness of the clan system which prevents the emergence of such a leader. As one informant put it (Pov Lis, Nomya, August 1982):

Clans exist so that the children can get married. . . . If there were no Hmong clans marriage would become impossible. According to our history, the Hmong are not (really) Hmong at all. Once they were called Ywj. There was one woman pregnant with child, and it was this which gave rise to all the Hmong clans; before that there were none. That woman gave birth to a baby son within the placenta, and the placenta, cut with a knife, produced twelve children, and they exclaimed, 'That's lucky! (hmoov)'. It was only since then that we were called Hmong. And it was those twelve children which formed the twelve Hmong clans, because all the children born after that time took their fathers' names, forming the clans. The truth is that although all the Hmong were born from that one mother, people don't think about it, and so we are unable to love each other. When the Hmong remember this, we will all be able to love each other.

As for those kings who were going to rise up amongst us but didn't, I don't know about all that. I only know that once long ago the Hmong had a King in Mongolia. Hmong were numerous and had kings but after that, when the Chinese made war in earnest, the Hmong were unable to win, and the Chinese took the Hmong characters and burned them up; and this crime was the foundation of the Hmong. Afterwards the Hmong learned Chinese, and shared their King. Some went to live in Vietnam, but the Chinese came to kill them and they moved into Laos. The Chinese let them go. The educated ones, the ones who knew books, went to live on the mountains, but after they died the Hmong had no more writing. Those Hmong kings which have come up occasionally were all killed, so that now there is no King appearing beneath Heaven for the people of this world. There has to be study and learning beforehand. Then there will be cleverness and wisdom, and it will be fitting to become officials, but it must be done slowly, slowly. Hmong should help Hmong and we should send our children to study high, and then Hmong will be able to help Hmong. . . .

The truth is that we all live in different places and different countries. The Hmong will not, cannot agree together. If only Hmong loved one another, then we would know the fullness of Heaven. The major causes of disputes between us are fines arising out of marriages; Hmong having no learning, don't know the natural way of mutual love; disputes over land, and great disputes because of love and adultery. That is why you will only find two or three households in a single village. Anyone who wants to help the Hmong should write this 'history' in a book for all to read. Now we already have our own writing, so that people in different countries can communicate with each other, and now we can teach the Hmong to love and help each other, to excel, and know that originally we lived in China, but the Chinese killed the Hmong King so he died, and then they gave opium to the Hmong to plant and let them grow it for the Chinese to buy, so that Hmong grew opium every day and then began to crave it and became addicted to it. Anybody who was a little clever ended up craving opium so then there was nobody clever left, because once you have made a sleeping place for your relatives to visit and smoke opium, then all the clever ones are finished, and that was how the Hmong got stuck on opium. But we've got to study techniques, science, skill, in order to better ourselves. . . .

This refers to the origin myth of the clans, which is always linked to the customs of marriage and courting, and which we consider later (cf. Lemoine 1972b). It is very often, then, the clan system, and its assumed connection with relations of affinity rather than those of descent, which is seen as the cause for the fission of Hmong communities, and the lack of unity which prevents the emergence of a real 'king' like that of the Thai, or Chinese.

Opposition to the Thai State

The CPT, officially established in 1942, declared its commitment to armed struggle in 1961 (Morell and Chai-Anan Samudavanija 1981). Following a military coup in Thailand in 1976, which ended a three-year period of parliamentary democracy, over 3,000 intellectuals had fled into the countryside to take refuge with the CPT. After Vietnam's 1979 invasion of Kampuchea and China's punitive invasion into Vietnam, serious conflicts began to develop within the CPT, and at the time of fieldwork, although heavy fighting still continued between CPT supporters and Thai government forces, mass surrenders and defections from the CPT were taking place under the terms of an amnesty offered by the government of Kriangsak Chomanand in 1978–9. Contacts had first been formed between Hmong in Thailand and members of the Pathet Lao in Laos during the early 1960s, owing to disputes with local police regarding poppy cultivation. In the early 1960s, rumours of the birth of a Hmong 'king' began to be disseminated which attracted many whole Hmong villages to move to guerrilla areas in North Chiangrai province. It was not until the beginning of 1967 that armed clashes broke out in Nan province.

It is fair to point out that most Hmong in Thailand have not been top-ranking CPT cadres, nor even particularly strong supporters of the CPT. The average Hmong village in Thailand remained, at this time, part of the Thai state, in that it had been affected by Thai developmental or educational innovations, and its trading networks were inextricably related to the main Thai market centres, although its villagers might know or have relatives among CPT supporters. Loyalties remained primarily towards other Hmong, through networks of kinship and affinal relationships, and extreme political polarization had not yet occurred. The fields cultivated, the acting networks of the lineage, and the ancestral rituals of descent groups were far more important to most Hmong than political commitments. The Hmong have thus been in the position of a client for whose support opposed political bodies have vied, and I found personal and idiosyncratic motives for political participation remained much more important than firm, ideological beliefs.

Most Hmong were fundamentally neutral with regard to the affairs of the Thai state, since they were to a large extent able to continue their traditional modes of subsistence. In extreme cases of conflict or trouble with the government, however, CPT-controlled areas provided an alternative sanctuary. Moreover, since clan loyalty took precedence over political affiliation, a Hmong visited by a lineage brother from the CPT would usually be expected to provide food and hospitality. Most of the Hmong I interviewed had joined the CPT to avoid economic ruin, or to escape for some reason from the Thai authorities. Many later found themselves disillusioned by encountering within the CPT the same sort of ethnic discrimination which they had suffered from before joining the CPT.

The following anecdote, from a household in the larger settlement, typifies the kind of reasons for which Hmong tended to move into CPT areas. Nyiaj Ntxawg had lived in Hapo since 1968, and had been born in 1948. Shortly after the Japanese retreat, which he remembered his father describing, his father's elder brother had died of smallpox, and a Karen had stolen some silver from his father's house. His father's eldest brother caught the Karen, and made him take him to the place in the forest where the Karen had hidden the silver. However, the thief had attacked him, and in the ensuing struggle Nyiaj Ntxawg's uncle had killed the thief accidentally with his sword. Nyiaj Ntxawg's uncle was put into jail for ten years as a result. He became extremely ill in jail, and had died within a year of his release. It was difficult for Nyiaj Ntxawg's father, deprived of the support of his two siblings, to make a living. It is in such situations that the CPT has provided an alternative.

We have seen how it was after several years of intensified government intervention in the Hmong economic system, that many Hmong originally fled to safer CPT zones. As in the case described above, it is largely out of the *economic* dilemmas into which the Hmong have been forced that their *political* dilemma has emerged. The attempt to maintain neutrality, however, in the face of what during the period of research was increased confrontation between the two parties, has increased the tension and uncertainty of the average villager's existence. One returnee to the survey site gave the following account:

Thirteen of us came back with nothing. I myself had not been as far as China, only as far as Laos. In Laos I saw nothing. We only saw Thai students from Bangkok there. Some of them had been to China and had come back to struggle for the towns from the forest. They told us they were only Thai students—from the south, the centre, the north, and the north-east. They came back after studying in China saying we should arise and make war—Thailand was too full of oppressors to live there any more. They went to the forest because two or three thousand people had been killed by the officials. They were very disturbed, and did not wish to live in the towns any more. If they had stayed in the towns they would have been killed with no chance to protest, so they had fled to the forest, so that if they were killed, at least they would be killed for [the sake of] the country. That is the way they talked. If we made war, once the land was secured we would no longer be looked down upon. We could become officials. Everybody would share their food and drink with everyone else, and help each other with clothing. And so everybody believed. For a long time we made war, but in the end it was not good, it was not as they had said. We had so many problems for living. We had no money and no clothes, no rice and no food. Whatever we grew they bought from us at the lowest prices: they paid 20 baht for a bip of rice. We could not live like that, so we came out. We stayed with them for ten years, and at the end we had nothing. We had only enough to get out with, and when that was finished, everything was finished. However many silver bars one had had in the beginning, they were all used up. Now we are very poor. They told us there was a king, with long ears which reached to his shoulders, and big eyes, so we went to see. But wherever we went we saw nothing but people like ourselves. They talked as if there was a King [*Vaj*] for us Hmong.[1]

If you had money [silver] you could leave, if you had none you could not. I am so happy to have been able to leave. If anyone ever comes here to deceive us again, I will not go with him, since there is nothing good in it. In the forest there is malaria, living in a land of war, great sickness. Some people were shot and died, others were bombed. I am happy to have been able to come out at last. I could not leave before: if I could have done, I would have. Those who went to China could see no *Huab Tais* [Emperor] it was all lies. They had said a *Huab Tais* had arisen in Laos, and there I searched for two or three years without seeing anything but people like us. They were just ordinary people who ate rice like you and me. No King eats rice like you and me. They were only leaders of people, like the Thais have a King, because he is clever. The students said they fled into the forest because so many were killed in Bangkok. Now they do not live in the forest: they have all returned. The government called them back, to be good people and study further and not shoot each other, but to go to Chiangmai and Bangkok.

I had asked this informant about the rumours which he claimed were circulated by (Hmong) Pathet Lao cadres in the 1960s that a Hmong King (*Vaj*) or Emperor (*Huab Tais*) had been born, and was summoning all his followers to a cave in Amphur Theung, in Chiangrai province. It was said that a Hmong King had been born to save the Hmong from suffering and misfortune. In the Hmong King's cave, everybody would be equal, and nobody would have to work for a living anymore. Each grain of rice would become a thousand grains, and there would be silver and gold for everyone. Many Hmong who believed these messianic rumours and went looking for their source, however, found themselves in a CPT base area. Messianic Buddhist uprisings, characterized by similar beliefs, have not been un-

common in Thailand. Almost invariably they have involved elements of
ethnic minorities (see Koch 1981; Tanabe 1984; Bunnag 1967; Keyes 1977).
Although Hmong messianism has absorbed elements both of Christianity
and of Thai or Lao Buddhism (for example, in the long ear-lobes attributed
to the Hmong King, which are one of the signs of the future Buddha), it
owes more to the long history of messianic and millenarian Taoist and
Buddhist popular rebellions in China (Wakeman 1977; F.-L. Davis 1977).
The essential connection between themes of rebellion and sovereignty
among the Hmong is a phenomenon towards which this book is moving.
At the same time, the strength of underlying resistance to the state expressed
through sporadic messianic or revolutionary movements should not be
obscured by the latter.[2]

Although the references to notions of the equal distribution of food and
clothing and the pervasiveness of oppression in Thailand are of an idealized
kind, they clearly refer to concretely felt and experienced phenomena,
such as the simple difficulties of making enough to eat, becoming officials,
and not being looked down upon.[3] Perhaps these are everybody's ideals.
Yet disillusionment with the CPT occurred just as much owing to the
actual calamities of living, as to feelings which much have been obvious
that a losing battle was being fought. The extent to which the motives for
such political participation remain of an *instrumental* kind is revealed by
the naïvety of the account above, where the idealized notions of the equal
distribution of food and clothing and complaints of oppression in Thailand
are based on concretely experienced phenomena, such as the actual difficulties of making enough to eat, becoming officials, and not being des-
pised. On the other hand, it is not surprising that disillusionment should
have occurred after so many years of guerrilla warfare. Living in CPT
areas, many Hmong had been unable to sell their crops or buy medicine,
salt, or cloth. For long periods they had been unable to light fires at night
for fear of being discovered, so that they could neither cook nor keep
warm on the mountains.

After surrendering his rifle at the BPP office in Theung and signing an
amnesty form, where the atmosphere was polite if somewhat strained,
Tsheej, the returnee quoted earlier, returned to a good position in the
village. At first he and his wife and four young children moved into his
elder brother's house, where the two families worked together for several
months clearing his elder brother's fields and new ones for Tsheej's family.
There was not enough room in the house for two families, however, and
relations between the two families were cool. Four months later, Tsheej
began to build a new (bamboo) house for himself and his family, at the far
end of the village. His brother helped him, and organized a collection of
rice for the new family from all the other Vaj households in the village. In
this way, the maintenance of traditional lineage reciprocity tends to prevent
the emergence of class stratifications in the village. Tsheej remained,
however, deeply scarred by his experiences with the CPT. I talked to him
many times, and it seemed to me that in some ways he felt superior to the
other villagers, and was resentful at having to accept his brother's and
their help. He had travelled further than they had, seen more things, and

received a broader education. Certainly the returnees, in general, spoke much better Thai than those who had remained in their villages and had a different appreciation of the Thai context. We see that the motivation for political participation among the Hmong against legitimate authority has remained largely instrumental, although also involving deferential elements, while support for legitimate authority was largely of a deferential kind, although inevitably elements of instrumentalism were also involved.

The Articulation of Conflict

As can be seen more clearly in the case of the refugees from Laos, already such experiences are passing into the realm of popular imagination and attempts are being made to reformulate them in terms of elements of Hmong folk culture. Below I give an example of a type of tale known as *dab neeg*, popular stories which usually refer to a long-distant time when the spiritual world of *yeeb* (yin) had not yet become separate from the mortal world of *yaj* (yang), so that men and spirits could communicate with one another. These *dab neeg* form some of the primary source materials in the remainder of this book. Through intertwining topical events into the fabric of a legendary past, the account below, told by a visitor who had come to Hapo looking for a girl to marry, reflects the transitional nature of the village as a whole (Hapo, January 1982):

The Story of Tsav Leej

There was one man named Tsav Leej, from Mae Chaem. For four days and four months he went into the spirit country, and yet he is still alive. He spent twelve months in Bangkhwen jail because his sister's husband kept going between Laos and Thailand (i.e., a communist) and the Border Police found many guns stored in his house, so they arrested him and took him to Bangkhwen jail. When he was released, they took him from Bangkhwen to Lampang by bus, but he disappeared, so his son, who had gone to Bangkhwen to meet him, was unable to find him. The son reported him lost to the police, who accompanied him back to Bangkhwen again but were unable to trace him, so he returned to Chiangmai and sent a message from the radio station there to say that his father was missing and if anyone knew of his whereabouts, to report it so that he could come to meet him.

But his father had a spirit friend, who had led him up the Mekong River, all the way to China and past China to the country of the spirits. For four months he went into the spirit country, staying on the bridge where people and spirits come to trade with one another, but none know which are spirits and which are people. People who wish to return must place some silver in the pail of water; if it sinks they may return, but if it does not then they cannot.

One day, after he had been gone some time, his spirit friend said, 'Tomorrow I will ask them [the spirit family with whom they were staying] to cook breakfast very early, and after breakfast I will take you right inside to visit the spirit country.' The next day after breakfast they went to meet the leader of the spirits there, to see what kind of licence they would issue him, of entry into the cold and dark country. So he replied that his uncle (FyB) had already died a long time before, and had a motor-cycle which could carry him. Then his uncle pleaded with the head of the spirits to write his nephew a card for him to go inside the cold and dark country. 'I have been here a long time already,' he begged, 'and I should very much like to see him.'

So the head of the spirits told Tsav Leej to swap clothes with his spirit friend, and at midday his uncle would come to collect him. 'You must return quickly,' he warned him, 'and if you are not back by midnight, I shall fine you.'

While Tsav Leej had been in prison his wife had remained alive, but she had died since then, and now he was able to see her. The Christian God lives up there too with the head of the spirits; they speak with the same voice.

At midday his uncle came to the bridge, along the heavenly way, and took him into the spirit country, where he met many spirits who asked him where he had come from. 'From Thailand,' he said. 'War in the world of people and war in the world of spirits,' said one among them, 'so you can go no further into the cold and dark country. It is because you are a good and honest person that you have been able to come so far to this place in the land of spirits.' 'We have heard of Thailand but never been there,' said the host of spirits around them: 'This way goes to France, and this way goes to Australia, and this way goes to England; all countries have ways to all countries, but this way you should not take: it leads through a hole to America. Come and visit our countries,' they all begged him. 'But I have already stayed a long time,' said Tsav Leej. 'I am not free to come and visit your countries.' 'If you come with us, we will look after you and you will be well,' they promised him. 'There are only five stages to our countries, where there are aeroplanes and cars.' But he and his spirit friend said they had already stayed too long, and could not go with them.

'There is war in both the world of spirits and the world of people, and only those like kings can go.' they replied. 'If you do not come further into our countries but return, both the land of spirits and the land of people will have war until brother will kill brother, husband and wife will kill one another, and the new sky will come.'

His friend the spirit took him back to his home in Chiangmai. On the way he lost sight of his friend, and so arrived alone. He did not know where he was, and thought that he might have died. But looking around he remembered the mountains, and remembered that the road was not far from where he stood.

So he fumbled his way to the road and took a car and came back: he arrived at night in his home, and his son sent a message to the radio station in Chiangmai saying his father had returned. Since he has come back he has said that brothers should not fight brothers and nor should husbands and wives; it was only like that with the communists, or else it would be as his friend in the spirit country had prophesied.

The story warns against the threat of political polarization. What is most striking about these types of stories, however, is how current events are blended with elements of a mythological past. I believe that this is because the Hmong Otherworld is so closely modelled on this world. The supernatural Otherworld supplies an inverted image of the social Otherworld represented in the past by the official Chinese state bureaucracy, and now to some extent, by Thai bureaucracy, replete with images of punishment and guilt, and characterized by essentially foreign demonic forces which must be placated or appeased. The boundary between ethnic groups, then, becomes transformed into the boundary between the natural and the supernatural realms. The *meeting place* between these two realms is symbolized by images of exchange and substitution which derive from the historic position of the Chinese *market* at the boundaries of the Hmong

world of mountain and forest. The Head of the Spirits is known as 'Nyuj
Vaj Tuam Teem', who judges and collects taxes in the afterlife. He is said
to be *tus plaub*, the Arbitrator, and his deputy is known as 'Ntxwj Nyug',
who issues licences for life which determine different individuals' life-
spans.[4]

This is a remarkable example of what Coleridge (1956) described as the
'synthetic, shaping' power of the 'Imagination', which blends and fuses
disparate elements; a 'modifying, esemplastic' power, which he distinguished from the 'aggregating' faculty of the 'Fancy'. We may see it as a
type of 'bricolage' which manipulates elements of the traditional belief
system in order to make sense of current actualities, and I examine this
notion further in the concept of 'real history' below. Meanwhile, I hope
that here I have succeeded in conveying something of what the conflict
between the Thai government and the CPT has meant to the Hmong, how
the basic alternatives of insurrection or loyalty to the Thai state are determined by essentially instrumentalist interests which arise from the economic dilemmas described in the preceding chapter, and how a typical
Hmong village may be defined and indeed define itself in terms of such
alternatives.

Neutrality

Many CPT-controlled areas in Thailand at this time were composed of
large intermediate areas of mountainside into which it was simply not safe
for outsiders to venture. Ethnic minorities in these areas to a large extent
continued to practise their customary modes of subsistence and belief in
return for defending the area and providing material support to CPT bases.
In this sense, the village maintained a real autonomy, and ideological
commitment was only expected at the extremities of the situation—where,
for example, BPP schools taught Hmong children loyalty to the King, or
where Hmong were educated in the history of colonialism in CPT bases.

In one large Hmong settlement to which I conducted a field-trip, although
it was nominally still under the supervision of the local Thai government
administration, all the peaks of the surrounding mountains constituted
a CPT-controlled area where many Hmong who had previously lived in
the settlement, or had relatives there, were located. Often Hmong from the
surrounding villages in the mountains would come secretly down to the
settlement to beg blankets, rice, oil for lamps, or ammunition from their
relatives in the settlement.

The following account by a Catholic missionary to the Hmong, well
expresses the essential powerlessness of many Hmong in the ideological
struggles of the Thai state (Mottin 1980, p. 62):

In 1974 we met Lao Nia Kao, who had surrendered two days before. He had never
been a communist, he explained. He had never shot at anybody and had not even
any arms. The attacks of 1968 had surprised him in the forest, but the very place
where he lived was still peaceful, so he had remained there. That lasted for six
years. The communists patrolled the region, and they met occasionally. Every

second month they were visited by two of them who chatted for a few minutes and gave them some medicine, generally out of date. Life became harder and harder. They were afraid to make too big fields for fear of being seen by the helicopters. They could not go to the market. They had neither salt nor cloth. Their clothes were falling into rags, and for two years, 'his trousers had a hole in the seat'. He no longer cut his hair. He decided to surrender.

Many Hmong remained in the deep forests which were also the province of Thai insurgents, and thus inevitably fell under their control. Certainly, ideological disputes regarding the Thai state had made themselves felt at the village level, but these were rather phrased in terms of a clash between Thai and Hmong interests. One young married man, for example, had made another girl pregnant. Since men are allowed to take second wives, and often marriage does not take place until pregnancy is assured, this was not adulterous behaviour in Hmong terms. However, a pregnant girl has the right to expect the man who made her pregnant to marry her, or else to demand a large fine, often in silver. In this case the girl was insisting that he marry her, and her family had demanded that he pay a very large fine, of seven silver bars, which was far beyond what he could afford, if he did not. However, he did not want to marry the girl, and, moreover, his wife had threatened to commit suicide if he did so, and his widowed mother was supporting his wife. Normally the affair would have been settled by negotiation between the two families or their descent groups through the mediation of elders and friends. In fact, his wife had no customary right to prevent him taking a second wife. However, the girl might have been persuaded to withdraw her claim or at least to reduce the fine demanded. But in this case the young man appealed to Thai law, rather than what the Thai called *prapayni chao khao* (the customs of the hillpeople), since Thai law would have demanded a much reduced fine. The local Mien teacher supported his claim. But he did in the end have to marry the girl, and eventually his first wife left him.

Although we can see here that consensual Hmong political solutions had been challenged to some extent by Thai law, the legitimacy of that law was itself under challenge at this time by other sectors of the Thai state, and it was such broader cleavages which constituted the major political alternatives for the average Hmong village. While neutrality in the conflict between the Thai government and its opponents was becoming more a matter of informed compromise, so too had the uncertainty deepened, of not knowing which way to turn, or who to believe. As uncertainty increased, so too did the anxiety which attended daily decision-making processes, much as it did in regard to the choice of whether to cultivate poppy or not. It was by such alternatives that the village was defined, and defined itself.

1. *Vaj* (or *Vaaj* in Green Hmong) corresponds to Chinese *wang*. *Huab Tais* corresponds to Chinese *Huang Ti*; although a Hmong variant, *Faj Tim*, would be the more logical variant phonologically (personal communication, Professor G. Downer, March 1984). Many stories are told of Hmong *Huab Tais* in different contexts. However, messianic prophets seem to see themselves as invariably messengers of the *Huab Tais*.

2. See the essays in Tanabe and Turton (1984).

3. Such data amply confirms the hypothesis that the 'redemptive process bears significantly on the politico-economic process' (Burridge 1971, p. 13).

4. Lemoine (1972b) identifies *Ntxwj Nyug* as *Yu Houang*, the Jade Emperor of the Taoist pantheon, who has here become identified with the Hindu–Buddhist *Yama* (*Yen Wang* in Chinese), the Lord of the Dead. The first two elements of *Nyuj Vaj Tuam Teem* he also identifies with *Yu Houang*, the Jade Emperor. The latter two derive from *ta tien*, 'great palace' (Lemoine 1972b), or perhaps 'great dynasty'. Although deities are often poetically doubled in the Hmong tradition, as Lemoine points out, it seems odd that *Nyuj Vaj* should not refer to the Chinese *Niu Wang*, or Ox-King (Werner 1932).

4

The Religious Dilemma

IN this chapter, we examine the effects of and motivations for conversion of the Hmong to the two most important religious alternatives which have been offered to them: Buddhism and Christianity. We find that, while the Buddhism presented to uplanders does not represent a genuine faith, but is rather seen as the ideology of a dominant ethnic group, Christianity has had a long history among the Hmong and has met with a certain degree of acceptance. However, Christianity offers to its minority converts an ideology which *transcends* the primary alternatives of assimilation to a Thai identity or the retention of a sense of Hmong ethnic identity, and thus properly speaking represents a third, or resolving, alternative to the possibilities of the continuation of pantheistic Hmong shamanism or the adoption of the state religion of Thailand.[1] Thus, the adoption of Christianity increases the conceptual distance between members of ethnic minorities and the states to which they belong. In considering this, we also examine the importance of the benefits of medical attention and literacy which Christianity has offered to the Hmong, and the association of Protestant Christianity with uprisings against majority populations of a millenarian and messianic nature.

Missionary Buddhism

Prior to the present century, distinctive local traditions of Buddhism had survived which were hierarchalized only to a limited extent. In 1902, King Chulalongkhorn centralized the Sangha, or monastic Order, which became accountable to a Supreme Patriarch, and growing state control of the clergy was reinforced by the Sangha Administration Act of 1963. As remote areas and the ethnic minorities who inhabited them became a security concern in the 1960s, the Thammacarik project of missionary Buddhism to upland areas was established in 1965 with the objectives of 'strengthening sentimental ties with the mountain people and creating loyalty to the nation by encouraging Buddhism'. The project formed the fourth 'Hilltribe Relations' branch of Thai government programmes for hill-dwelling ethnic minorities, and was an offshoot of a broader Thammathud programme of harnessing Buddhism to developmental measures.

The missionary monks have attached particular importance to demonstrating to the uplanders how monks behave and teaching them how to make obeisances to the 'Triple Gem' of Buddhism (the Buddha, his Teaching, and his Clergy). They have also stressed the importance of communicating one of the fundamental concepts of Thai Buddhism; that by good deeds, such as offering alms to the monks or oneself ordaining as a monk or novice monk, one can 'make merit' for oneself which will ensure a fortunate future. This ritualistic type of Buddhism is one deeply embedded in the context of Thai society. Also it relies on a distinctly modernist interpretation of the self-abnegating philosophy of the Buddha, closely associated with the legitimation of the Thai state. Thus it has tended to be seen by the ethnic minorities as the ideology of a dominant ethnic group, and they have very largely failed to adopt it.

The Thammacarik programme has recruited monks from temples in Bangkok and Chiangmai to visit hill stations for varying periods. It has constructed temples in upland areas, and ordained a number of minority monks, some of whom have, in turn, themselves become Thammacarik monks.

Keyes (1971) found that very few of these newly ordained monks remained ordained even for short periods of time, either because of difficulties in supporting their families while remaining monks, or because of their inability to speak or write Thai. There was a tendency to despatch the members of one ethnic minority to preach in villages belonging to other ethnic minorities, where the languages they spoke could not be understood, and in one area Keyes documented an almost complete lack of communication between the monks and local people.

In Nomya, the headman's second son ordained in a Chiangmai temple, and he did so explicitly in order to be able to learn how to read and write in Thai. He complained to his father that he was suffering greatly from being given only one meal a day, and his attempts did not last long. In Hapo, there was only one villager who claimed to be a Buddhist, and this claim was little more than a symbolic statement expressing his identification with the Thai state. Although he wore a Buddhist medallion around his neck, and had spent two years in a Buddhist temple as part of his education, he showed little knowledge of the most basic Buddhist beliefs or rituals.

Yis was a man of twenty-two, who lived with his wife and his widowed mother. His father had died when he was seven, and he had had to support his family from an early age. He had done extremely well, particularly through forming contacts with local Thai interests; he spoke better Thai than the other villagers, and took it upon himself to act as an intermediary between other villagers and the Thai teachers. He and his family had worked so hard in previous years that they had become self-sufficient in rice in 1980, and Yis could afford, at a very early age, not to do very much, but leave most of the work of tending the fields to his wife and his mother (who ruled the household). He was a clever, impulsive person, who was in serious difficulties with his wife over a girl-friend. His wife had threatened to leave him if he took his girl-friend for a second wife, and the girl-friend's family was demanding a large fine for the child she claimed was his unless he married her (a similar case to the one outlined in Nomya, page 83).

His mother was a particularly hard-working, shrewd woman, who had sent Yis off to the temple at an early age, and together with his wife, worked the fields which supported the family. Although I often spoke to Yis about his Buddhism, I was never able to get very far. He would invite Hmong shamans to perform shamanic rituals if any member of his family was ill, and although he did not practise shamanism (*ua neeb*) himself, he performed household ceremonies and sacrificed to the ancestral spirits (*ua dab*) like all other heads of households.

Yet, when the nearest Thai temple held its annual *kathin* ceremony at the end of the Buddhist Lent, when new robes are presented to the monks by villagers, Yis organized the group of five Hmong villagers who went to watch the ceremony. Curiosity and not devotion was the major motivation here; several Hmong attended a local Thai female spirit-medium festival at the mining settlement, and on another occasion in Chiangmai watched in silent, fascinated horror the devouring of raw, bloody meat by a female medium. This was something the Hmong could never do.

In a questionnaire I distributed among Hmong who had settled in Chiangmai town, I found thirty-three out of a total of 128 who were attached to three separate Buddhist monasteries. However, not one of these claimed that their religion was Buddhist. Thirty-one declared that their purpose in joining a Buddhist monastery was to study, and the remaining two stated they had been sent by their fathers, one of them to study. The majority had been able to reach levels three to four of secondary education in this way. Of the others, one nineteen-year-old was in adult education, and the other was taking a bachelor's degree in medicine at Chiangmai University, while staying in the temple. In fact, only twenty-five of the remaining sample were not involved in some way with education, and some of those who were not had children who were at school, for whose benefit they remained in town. In the primary value accorded to education by the thirty-three at Buddhist monasteries, they differed little from the majority of (Thai) Buddhist novices, many of whom join the Sangha on account of social pressure and the mobility such a step can lead to, or because monkhood enables a course of literacy to be pursued.[2] The great importance of acquiring literacy for the Hmong is examined in Chapter 6 below. The Hmong recruits were quite frank about this, and not one of them considered himself to be Buddhist.

In the survey site, there were no religious establishments of any type. Hmong worship generally takes place in the household and does not require the establishment of special churches or temples. Under the Thammacarik programme, however, a monastic temple had been established above the nearby mining concession, with at any time between two and three monks resident there. The Abbot of this temple, who often visited the local villages to preach, originated from southern Thailand and had difficulties in understanding the local dialect. He was notorious in the district, since it was said that previously he had been involved in the prostitution business, which was prevalent in the mining settlements. He had become a monk only because a local (female) Thai *wessa* (wizard) had prophesied that he would die unless he repented (cf. Khinthitsa 1983).[3]

While his status as a monk may have redeemed his previous behaviour in Thai terms, it did not do so for the Hmong, who were unable to respect him as a spiritual leader. In my conversations with him, he invariably saw his task as one of bringing development to the poor, primitive peoples who constituted the minorities of the hills.

The Conversion of Tua

It is because the type of Buddhism presented to the Hmong and other ethnic minorities is so closely associated with the fundamental values and orientations of Thai society that it has largely failed to be adopted by them in any widespread or meaningful sense. One extraordinary fictionalized account of a Hmong conversion to Thai Buddhist, penned by a famous Thai Buddhist monk, well illustrates the clash of value systems involved (Dhamaraso and Virojano 1973). The story tells how a handsome young Hmong hunter, named Tua, wishes to relieve himself in the forest one day, but is shocked to discover a beautiful young girl named Mua already squatting at a tree trunk, completely naked. She does not notice him because she is absorbed in the rare sight of two chameleons mating, and Tua is too astonished at his first sight of a naked female body to move. When he recovers his senses he shouts 'Bang', frightening the girl so much that she runs away 'without even cleaning herself'. Tua thinks constantly about the girl, and eventually they meet again and begin to fall in love. However, Tua is so impatient that he arranges for two of his friends to help kidnap her.

Late at night, Tua shines a light into her parents' house to let Mua know he is there, and she steals out to join him. When she sees his two friends, who seize hold of her, however, she screams out and her mother and relatives come to the rescue. After a fierce battle, marked by kicking, punching, slapping, and spitting, Tua wins his bride and she comes under the protection of his mother, who uses a hen to cleanse her of bad spirits. According to the story, she was then accepted as Tua's bride 'against her will' and had to stay in the house for three full days without going anywhere else. The day after her capture, her parents come to retrieve her, but fail. For two years, however, Tua and Mua have no children—until one day, walking in the forest, he meets a Buddhist monk. Impressed by the monk's air of contentedness and fortitude, Tua seeks instruction from him and agrees to observe the Five Precepts against killing, stealing, illicit sex, lying, and alcohol. When he returns home that night, Mua, his wife, is afraid that he will die if he breaks one of the Precepts, and persuades him to go back to the monk and recant. The monk, however, persuades him at least to keep the Third Precept, which forbids adultery, and this delights his wife. Finally, Tua bags a large deer which he decides to take to town to sell.

Along the way, however, there lives the faithless young bride of a sixty-five-year-old millionaire, who is pining for a lover. Little does she know that her rich husband suspects her, and has devised a trick to test her fidelity. Thinking her husband has left for another province, she opens her window from the third floor of her house and spies the handsome young tribesman walking nimbly along the road: 'She saw a man dressed in black, wearing a black cap with a red spot on top. Around his collar were five or six strings of beautiful, glittering beads. His blouse was embroidered with flower patterns of alternating colours, truly beautiful to behold.'

Impetuously she despatches her servant girl to invite Tua to stay. He is tired and wishes to escape the heat of the town's cement buildings, so he agrees. He is served splendid food which he does not know how to eat, and late at night the millionaire's wife comes to his room to seduce him, smiling seductively and wearing a see-through sleeping-suit. She is heavily powdered and perfumed, and her unnaturally red lips are compared to the lips of an ogress who has just drunk blood. Tua cannot sleep with her because he has agreed to observe the Precept forbidding adultery: 'I am a married man,' he exclaims. 'Your wife must be very beautiful, I suppose.'—'No, she is also a tribeswoman.'

So she threatens to accuse him of rape and have him put to death if he resists her advances but Tua remains adamant. The millionaire reveals himself, banishes his wife, and invites Tua and Mua to live with him and collect his rent for him. They are immensely grateful, since this will relieve them of the hardship of having to rely on hunting for a living. As a result of his faithfulness to the Buddhist Precept, Tua and Mua live happily ever after, as Buddhists. He offers food to the monk every day, and has a temple built to express his gratitude for his conversion.

This fable displays a surprising familiarity with certain details of Hmong custom which are skilfully employed to drive home the author's main point about what constitutes proper sexual behaviour, which is, of course, the major topic of the tale, although there is a minor theme related to the taking of life forbidden by the First Precept. Neither the behaviour of the unfaithful Thai wife nor Tua's capture of his bride qualify as 'proper sexual conduct' in the sense in which the Third Precept is understood in Thailand. However, when the author describes the capture of the bride a contradiction appears, for the account simultaneously appears to suggest that Mua was willing to follow Tua to his home while later declaring that she became his wife 'against her will'.

In fact, the type of marriage described by the author is customary behaviour among the Hmong, although he nowhere suggests this, and the account is presented in such detail, like the account of naked people relieving themselves in the forest, as to shock a Thai readership at such aberrant sexual behaviour among the Hmong. Marriage by capture is a formalized manner of gaining a bride in certain situations, such as when the girl's parents disapprove of the marriage, and it usually takes place with the fore-knowledge of the bride (as the account in part suggests). Moreover, it must be performed correctly; it should take place after sunset, on an even rather than an odd (that is, female or waxing rather than male or waning) day and month of the lunar calendar, and it should take place more than three double armspans away from the girl's house. It is not the kind of unsanctioned behaviour which the author presents it as. Whether the author was aware of this, it is difficult to say, since he might be merely repeating a verbatim account of such a marriage without realizing its social significance. But I suspect that he does fully recognize the implications of such marriages, since he includes other details, such as the ritual waving of a rooster three times around a bride's head at the threshold of her new home to banish her ancestral clan spirits which also demonstrate a familiarity with Hmong customs. The story is addressed on different levels to both a Hmong and a Thai readership. While employing the details of a marriage by capture to antagonize a Thai readership, the author, at the same time, employs the detail of the couple watching the chameleons mate to repel a Hmong readership. For this is a classic example, in terms of Hmong belief, of a situation in which soul-loss, leading to illness, will occur. It is rigorously prohibited to watch animals couple, since the soul-substance of the observer will be attracted out of the body of the observer and absorbed into the observed womb to be reborn as, in this case, a chameleon. Without the soul-substance, the body of the observer will sicken and may die. A Hmong reader would thus understand that Mua must have fallen gravely ill after such an experience, to the extent that she could not have children, and that this was probably why Tua had recourse to the Buddhist monk, who blessed the couple with happiness and prosperity.

The account of the seduction is a rather obvious male fantasy which its Buddhist author compensates for by painting the rich man's wife as an ogress of the kind often depicted on Thai temple murals. Yet the climax of the story comes in the juxtaposition of the free, independent, outlandish, and exotic Hmong 'tribesman' striding down the road as the wicked, corrupt millionaire's wife gazes out from the *third* floor of her evidently palatial mansion. Hmong innocence is further portrayed in Tua's unfamiliarity with modern amenities such as cutlery and electric lighting, and in the dialogue between Tua and the monk, where the simplicity of the Hmong as model acolyte contrasts powerfully with the magisterial utterances of the paternalistic monk.

Tua is originally presented to us as a hunter, and it is coherent in terms of the author's purpose that his ultimate happiness should depend on his surrendering this sort of livelihood. The fundamental incompatibility of Buddhism with the Hmong belief-system is, however, revealed in the Preface above, where the author suggests that it may be because of their sin in killing animals that the Hmong are so unhealthy. Certainly in Buddhist terms, transgression of the First Precept against taking life would lead to unwholesome consequences for the transgressor. Yet the Hmong normally only kill animals as sacrifices in order to cure illness, since illness is often attributed to the capture of human soul-substance by malevolent spirits, or *dab*, who may be persuaded by the shaman to exchange it for the soul-substance of a sacrificed animal.

As a whole, the account provides a most interesting illustration not only of the kind of ethnic stereotypes which the Thai may have about the Hmong, but also of the fundamental incompatibility between two quite different systems of values and beliefs, based on radically different social orders. It is this which, in large part, explains the failure of the Hmong to adopt Thai Buddhism under the Thammacarik programme of missionary

Buddhism. Since Thai Buddhism is so closely associated with the values and ideals of Thai society and the Thai nation state, the adoption of Buddhism implies assimilation into Thai society and the loss of Hmong ethnic identity. This, as we have seen, has not occurred.

Buddhism, then, impinged on the lives of the Hmong in the survey site in a number of ways. Primarily, it offered an opportunity of education, and this is the reason that some teenagers were sent to ordain for a number of years—not to learn Pali, the language of the Buddhist texts, but to learn Thai. To a large extent, Buddhism remained the creed of the ruling classes, to which Hmong might aspire, but could not, in terms of their own beliefs, take very seriously. In general, it remained an object of curiosity and rarity value, to be inspected, like the female mediums, from a discreet distance. Emblems of it might be adopted to exemplify loyalty to or sympathy with the Thai state, rather as posters of the King might be displayed, as in Yis' medallion. But overall, there remained the story of the local Thai monk, which everybody knew, and his deserted temple at the front of the mountain two valleys away.

Missionary Christianity: The Initial Impact

In this section, we consider some of the themes which are the subject matter of later chapters which deal with notions of literacy among the Hmong and ideals of messianic leadership, since both have been articulated through the medium of Christianity. In times of extreme economic deprivation, recourse has been had both to violent uprising and to the adoption of Christianity. Occasionally the two strategies have been adopted together. A historical account of the impact of Christianity on the Hmong is given here which pays particular attention to themes of literacy and themes of messianic rebellion against the state, and the section concludes with an analysis of some individual cases of the adoption of Christianity and villagers' views of Christianity from the survey site. The focal study of the village which follows draws out these themes and extends them in the light of the impact of the adoption of Christianity on the focal village as a whole.

Christian mission work among the Hmong has had a long and well-established ancestry. It seems often to have carried with it something of a mass appeal, particularly in situations of severe economic stress, and to have achieved much of its success through the prospects of education, and, above all, literacy, which it promised. Although Catholic missionary work may have begun earlier,[4] Protestant missionary work first began among the Hmong of the North-east Yunnan and North-west Guizhou areas shortly before the turn of the century.[5] In 1904, a group of Miao arrived in the city of Zhaotong seeking a Methodist missionary named Samuel Pollard, to whom they had been directed by a Mr Adam of the China Inland Mission at An-shun, who later translated the New Testament into a romanized version of Hmong.[6] Work which originated with the China Inland Mission was later carried on under the auspices of the Yunnan District Mission (Smith 1932). Pollard maintained a mission station at Zhaotong, twelve days north of Yunnanfu, but as greater and greater numbers of the Miao started arriving, causing trouble with the majority Chinese population of the city, he constructed a special Miao mission station with a school and church at a place known as 'Stone Gateway' in the hills of Guizhou, which had been donated by a rich converted Lolo landlord named An-yung Cher.[7] Over a thousand worshippers used to come to church there every Sunday (Hayes 1928). Other churches, dispensaries, and schools followed in outlying Miao districts.

Pollard campaigned for the land rights of the dispossessed Miao against the Lolo and Chinese landlords who exploited them, conducted mass baptisms in what the Chinese claimed were 'magic trunks' which gave the Miao supernatural powers, and invented the first romanized script for the language, in which the Bible was eventually printed. He initiated mass smallpox vaccinations and burned to the ground covered houses which he thought were used for courting, concentrating particularly on converting the 'wizards', and introduced alternative games, athletics, and competitions to replace traditional Miao festivals (Dyke 1964; Grist 1971; Moody 1956). By the time of his death in 1915, of typhoid fever, native trained teachers were in charge of seventeen branch schools, other missionaries had taken his place, and three years later the first Miao scholar headed the lists in Mathematics at the University of Chengdu (W. Pollard 1928).

At the time Pollard came across the Miao, they were in a desperate economic situation (well described by Pollard in the many writings he has left us) after the suppression of their last two great rebellions of 1797 and 1856. In 1800, 5,600 Miao were reported arriving in Yunnan from Guizhou, and the same year a major migration took place to Tongkin in Northern Vietnam, from where they penetrated Laos. Travellers through Guizhou in the last quarter of the nineteenth century described some of the devastation which these rebellions had left behind (Wingate 1940):

At a place called Ching-chi ... is a town wall of some miles in circumference which once enclosed a prosperous town of several thousand families, but which now contained only a few officials' houses and a temple or two. It is one among others which have suffered during the wars which the Chinese have had with the Miao or aborigines living in the hills south of the river.

Likewise Margary, before he was murdered prospecting for the great projected Yunnan–Burma railway, found whole deserted areas where the Miao had 'come down from the hills and butchered the whole population', and spoke of their suffering from 'scorn, contempt, and legal robbery in rents and taxes' (J. Anderson 1878).

These uprisings arose from systematic attempts by the Manchu rulers of China to dispossess the Miao and other inhabitants from the fertile forested regions they occupied, which were rich in products such as iron, copper, gold, silver, marble, and wood (Davies 1909, p. 310). Thus Chinese settlers were brought in under land campaigns to displace the minorities, whose customs and languages were prohibited (see Lombard-Salmon 1972).

This, then, was the situation into which the first Christian missionaries entered. Pollard describes the large percentages of rice or corn which had to be delivered personally by the Miao tenants to their feudal landlords, and the monthly measures of wine demanded of them. Persecution of the Miao seems to have been widespread (Kendall 1954):

The first night we stayed in a Miao house, more well-to-do than the ordinary Miao: the lower part was a cob wall and the upper mostly open or stopped up with straw and boards. Plenty of fresh air and plenty of draught. There was a crowd of people in the house, the old man, three sons, two married with families, five or six grandsons. Formerly he had been better off, but when they get out of poverty the covetous eyes of their Norsu neighbours fall on them. He was accused of stealing, and with one of his sons was led off in chains and tied up for a month. His thumb was pricked with red-hot tongs, and he was beaten on the back with a sword. That was nine months ago, and the marks are still there on his thumb and his arm is painful. His oxen, horses and sheep were driven away, a fine imposed, and then they were released. It appears that this kind of oppression is common. A Miao dare not get rich or his landlord will take away his wealth.

We have seen how the production of opium is particularly liable to lead to a situation of protective extortion, and it was probably at about this time that the Miao in China first began to cultivate the poppy commercially, although McCarthy had recorded it in 1888 for Laos (McCarthy 1900).

It was during the first decade of this century that China's own home campaign to eradicate opium production was initiated, and consequently poppy fields were moving away from the sides of villages and highways to the more remote regions inhabited by ethnic minorities.[8] Pollard describes how Chinese speculators were leasing land on Lolo territory for the cultivation of poppy which they tried to persuade them to undertake on their behalf. Many of the labourers on these fields must have been Miao, since they were the tenants of the Lolo, and Pollard's diaries contain many references to the absence of opium among the Miao at this time: 'Here there were idols and opium, a different world from the Miao,' he remarks of a Chinese house in 1904 (Kendall 1954, p. 139). The Miao were, at any rate, the target of extortions and oppression of every kind.

It was not surprising, then, that the appeal which Christianity made to these desperate people was a mass one. Pollard describes the original impact of the Gospel on them in the following words (S. Pollard 1919, p. 72):

Some days they came in tens and twenties! Some days in sixties and seventies! Then came a hundred! Then came two hundred! Three hundred! Four hundred! At last, on one special occasion, a thousand of these mountain men came in one day! When they came, the snow was on the ground, and terrible had been the snow on the hills they crossed over. What a great crowd it was!

And Hudspeth, Pollard's colleague and disciple, who continued his work with the Miao in Southern China during the 1930s, adds the following description (Hudspeth 1937, p. 89): 'Picture these early scouts as they journey. Each carries a felt cape which in the day-time serves as an overcoat and at night as a blanket, and every one has a bag of oatmeal, a basin and chopsticks or a wooden spoon; the oatmeal mixed with water serves as breakfast, lunch and evening meal.'[9]

Extraordinarily enough, however, what had attracted these great crowds was not initially economic succour, but a rumour that the missionary had a *book*, or *books*, which was specially meant for the Miao. Pollard describes how 'The great demand these crowds made was for *books*', and how the Miao converts flooded into his house, eager to devour anything of written matter with which the missionary could provide them, and rapidly studied to become literate in their own language, in the script which Pollard invented for them. Later we examine the great significance this promise of their own writing, to a people who had none, would have had for the Hmong, in terms of their own folk-tales and legends. Here it will suffice to point out that evidently the primary goal of these converts was for literacy, since the 'good book' had become identified, whether by accident or design it is impossible at this stage to say, with the Hmong books or writing which the Hmong claim to have *lost* owing to the encroachments of the Han Chinese upon them and their territories. Such identifications of Bibles and the missionaries who brought them with elements of ancient prophecies seem to have been particularly common in the region.[10]

But this was not the only misinterpretation of the message Pollard sought to convey which occurred. The impact of Christianity led directly to the emergence of messianic movements (S. Pollard 1919, p. 76):[11]

We were troubled in yet another way. By some means or other the rumour went abroad that Jesus was coming again very soon. Instead of the teachings of the Second Coming proving a blessing to these simple people, the way in which it was stated by some irresponsible and ignorant people led to disastrous results. Some of the old wizards, and some of the singing women tried the role of prophet, and several dates were announced for the appearance of Christ. So firmly did some of the people believe these prophecies, that they neglected their farm work and gave themselves up to singing and waiting for Jesus. One party betook themselves to a loft, and with lighted lamps or torches stayed up all night, expecting the King every moment. Poor, simple people, one cannot even smile at their misled enthusiasm. They had known the bitterness and degradation of heathenism so long, that one cannot wonder at their hoping for a short cut to the Millennium, when all wrongs would be righted and everybody have a chance.

Of a piece with this is Pollard's account of a Miao couple who suddenly became deranged and began shouting and dancing around the mission house, the 'mad Miao woman' claiming to be Jesus's sister (S. Pollard 1921, p. 45). As we show later, the relationship between Protestant Christianity and messianism among the Hmong has remained a constant one.

Faced with these extraordinary cases of possession, mass migrations, and literacy among the subjugated Miao, the Chinese population, already increasingly resentful of the position of foreigners in China after the suppression of the Boxer Rebellion of 1900, became alarmed, as did many of the Lolo landlords who were frightened of the prospect of their tenants' emancipation. Church membership gave the Miao a privileged status, and Pollard quotes a Lolo landlord as saying that no one now dared to molest the Miao as they had done formerly (S. Pollard 1919). Rumours began to circulate regarding Pollard among the Chinese. It was said that the foreigners were supplying the tribespeople with bags of poison with which to pollute Chinese and Lolo water supplies, and that Pollard, who could see 3 ft into the earth, was dropping magical water into Miao mouths which instantly enabled them to read Chinese characters and develop prodigious memories. The Chinese warned the Miao to keep away from the foreigners, or else the foreigners would take out their eyes (to which a Miao is said to have replied, 'Quite true, quite true, he has taken them out and changed them. With the old ones we could not read, but with the new ones we can' (Kendall 1954, p. 92). A Miao rebellion was feared, and converts visiting the city were beaten and turned away, tortured, and even imprisoned in dungeons. Pollard himself was set upon and nearly killed. It was for these reasons that Pollard eventually moved his centre of operations outside the city, and established 'Stone Gateway'. Clearly, the introduction of Christianity, to a minority people who were already marginalized in relation to their majority neighbours, the Chinese, had the effect of *widening* the conceptual distance between the ethnic groups rather than narrowing it, and, in effect, contributed to the further marginalization of the converts. This, I shall argue, has been the general effect of the impact of Christianity on the Hmong.

Missionary Christianity: Later Effects

In July 1918, the Hmong of Laos, suffering from the taxes levied on opium collected by local Tai officials appointed by the French on behalf of the Opium Monopoly, rose up in the last of their major rebellions, against the Tai, Lao, and French. This insurrection, which later became known among the Hmong as the 'Guerre du Fou', was of a serious nature, and took the authorities three full years to suppress. It was led by an orphan named Paj Cai, who was affiliated to the Vwj clan. Paj Cai claimed to be possessed by the spirit of the Hmong *Huab Tais*, or Emperor, and various magical supernatural powers were attributed to him. Savina, a Catholic missionary who worked with the Hmong throughout this time, has described how the Hmong presented themselves fearlessly before the rifles of the French, believing that the bullets would not leave their barrels, and how the young girl who accompanied the troops into battle would try to catch the bullets in her apron (Savina 1930, p. 258). Paj Cai, sometimes described as an epileptic, is said to have wished to establish a separatist Hmong kingdom, and the movement was joined by the members of other minorities. After the suppression of the revolt, administrative reforms were put into effect which granted the Hmong some measure of autonomy. There is evidence to suggest that the movement was influenced by Hmong from Yunnan (according to Savina (1924), its leader was maintained by the 'king' of the rebellious Miao in Yunnan), and indeed that it began from a movement within the frontiers of Yunnan which had been directed against the Chinese (Allton 1978). This is lent some credence in that it coincided with a severe drought and famine which was recorded in South-west China at the time (*Chinese Recorder and Missionary Journal* 1919, Vol. VI):

Famine conditions this last spring were not as bad among the tribespeople in Yunnan as in Kweichow, yet both Miao and Lisu in parts of the northwest were reduced to eating the roots of ferns, first pounding out the juice, then boiling it until the pulp turned black. Even this food could be had only in small quantities and contained so little nutriment that those feeding on it were soon too weak to climb the mountains.

Under the 'Miao Famine Fund' set up by Methodist and other missionaries, money was collected from England and distributed to the needy Miao, and a great number of conversions were made (*Chinese Recorder and Missionary Journal* 1919, Vol. VI):

Many thousands of aboriginals were for months face to face with death by starvation. The relief which came from England and from Shanghai and other parts of China saved the Flowery Miao tribe from partial extinction and dispersion. . . . The hearts of the people have been moved towards God as never before. Within three months we baptised over 1,800 persons, and many more are preparing for baptism this year.[12]

This is a classic situation in which messianism is likely to occur (Burridge 1971, p. 108), and it is striking that just such a messianic movement should have been in existence in neighbouring Laos. Certainly one can imagine the effect which large amounts of aid, for which no ordinary economic origins could be discerned, would have had upon a people at the point of starvation. A precisely similar situation was later to occur in Laos during the 1960s among Hmong refugees supported by USAID-sponsored airlifts of rice, which we consider later.

Further migrations of Hmong from China into Laos took place throughout this period. I strongly suspect that Paj Cai was influenced by Christian missionary teachings, or that many of his followers were. For he taught the following (Yaj 1972):[13]

Everybody should love their children, and after they have been born their hands and feet should be washed with hot water. Clothes must be taken care of. Do not let the mosquitoes bite at night: such a person will become weak or sick. As for meat, eat only the flesh of animals expressly killed for eating. Do not eat animals which have died in giving birth, nor of internal diseases (*hlav hnyuv*). Do not dig out the burrows of field-mice. Do not hunt birds or squirrels; each has one life, and does not wish to die. As for the trees and bamboos, take only those of which you have need. Those you do not need, do not cut them down [just] to be thrown away. From now on, do not kill many animals when someone dies. If you truly love your ancestors, sacrifice only one. If two or three are killed, they should only be for eating. And do not skin animals, but carve the skin together with the meat for the eating.[14] To skin animals only destroys their bodies for the sake of eating the meat; the spirit impaired in such a way will mount to heaven to register its complaint. And as there is one there who judges these cases, this will cause difficulties.

Certainly the reference to mosquitoes would seem to indicate missionary influence. However, typologically the movement should be distinguished from those messianic movements which took place completely within Christian frames of reference considered above, which we continue to summarize later, although we may see both alike as a common response to a fundamental situation of economic hardship (a point which Savina (1924) makes particularly strongly with regard to the Paj Cai revolt: '*C'est un miracle que ces Meos là ne se revoltent pas,'* Savina was told by a French colonial official *before* the onset of the revolt). Thus, it may have been the initial impact of Christianity on the indigenous belief system which provoked the emergence of a nativistic kind of messianism.

For Protestant work continued, with the Hmong of Indo-China, to provoke messianic movements of a Christian kind. Although Protestant missionaries had worked in Laos since about 1930, the first missionaries to contact the Hmong of Xieng Khoung province arrived there in 1940, and their work, which was mostly of a linguistic nature at this time, was interrupted by the outbreak of the Second World War, and the Japanese Occupation. In 1949, however, when missionary work resumed, nearly 1,000 Hmong converts were made in a single day. Eight months after this, the wife and family of the most prestigious Hmong in Laos at the time, Tub Npis Lis Foom (who was later to become Minister of Justice in Souvannaphouma's neutralist government of 1960) embraced Christianity, Tub Npis himself declining for fear of offending the Lao Buddhist authorities. The example of his family, however, encouraged further conversions among the Hmong (Smalley 1956):

For years the Christian witness made no impression on the Meo. Then suddenly, in the space of a month in 1949, about a thousand converts were made. Today there are several thousand Meo Christians. Furthermore, occasionally 'prophets' declare themselves to be Jesus. So far none of these splinter movements has become widespread, but they are symptomatic of the fact that the Meo are undergoing a period of cultural reformulation which was triggered and given its particular form by the Christian gospel.

These early conversions were inspired by the prophecy by a female shaman named Po Si of the imminent coming of the Hmong *Huab Tais* (Barney 1957).

By March 1957, when the Viet Minh occupation caused the evacuation of the entire province by the missionaries, some 3,000 converts had been made in fifty-six villages, rising to 5,000–6,000 in ninety-six villages, of whom about 70 per cent were Hmong (Barney 1957). Here, too, a movement known as the 'Meo Trinity cult' grew up: three Hmong, claiming to represent the Holy Trinity, travelled from village to village winning converts. They burned household altars as the missionaries did (and still do), they removed 'fetishes' from the body, such as the silver neck-rings often worn by Hmong, and they performed rituals to exorcize evil spirits. The movement came to an end when the 'Holy Spirit' killed himself by jumping off a high mountain ledge to prove that he could fly like a dove (Barney 1957).

Still later, during the escalation of the Indo-China conflict in the 1960s, when thousands of Hmong were forced to become refugees in the 'free-fire' zones and became dependent on airlifts of rice from CIA-chartered airlines, messianic myths pervaded the society: 'The rumour circulated rapidly among these credulous tribes that a prophet who had fallen from the sky was summoning all the Meos to revolt and form a great independent kingdom with its capital at Dien Bien Phu or Muong Heup' (Halpern 1964).

One such rumour reported was that Christ was about to appear to the Hmong in a jeep, wearing American clothes and handing out weapons (Garrett 1974).

Thus, there is a close connection between the adoption of Christian beliefs and the emergence of messianic beliefs which may be seen as a *response* to such adoption. This is brought out very clearly by a recent missionary account, with reference to the movement in Thailand during the early 1960s, which affected Hmong from the survey site, as described in the account of Tsheej given in Chapter 3. In 1960, these rumours of a Hmong 'King' throughout the Northern Thai hills, which drew 'several whole villages' to China in search of their source, caused the flight of many missionaries, since the 'Meo King' group had been told to 'get rid of all foreigners' (M. Heimbach 1976, p. 35). Missionary houses were burned, and the lives of Christian converts and missionaries were threatened. We may conclude from this that the messianic movements which have occurred among the Hmong fundamentally represent a reaction *against* missionary influence. Messianism is thus to be seen as the *response* of the traditional belief-system to alien belief-systems which challenge it. We will return to more indigenous forms of messianism in Chapter 7, and the association of the messianic *Huab Tais* with ideas of regaining the lost form of Hmong writing is examined in Chapter 6. Here one may note the appeal which Christianity has made to those ideals of regaining literacy, and conclude that the adoption of Christianity has never lessened the social distance between the minorities and the majorities with whom they, in many cases, maintained difficult and troubled relations before the advent of Christianity. Rather, the adoption of Christianity has usually alienated the minorities from the states of which they were, perforce, a part. The end results of this process of alienation may be seen in the case of the Hmong refugees from Laos.

Probably a third of the Hmong population of Laos was enticed or beguiled into the fold of the Royal Lao government, largely through the auspices of fundamentalist Christian aid workers, such as the notorious 'Pop Buell' (Schanke 1970). Here, too, the promise of literacy played a significant part in winning Hmong support. Schanke, who portrays Pop Buell himself as a messianic hero, reports the following conversation between 'Pop' and a Hmong: '"You will build the school and operate it, Tan Pop?" "No, I won't do that. But I'll help you help yourselves, and I'll see that you get enough supplies to keep it going. America can do that much for you"' (Schanke 1970, p. 84).

Today, these Hmong have been welcomed into new homes in the United States, often by Christian pastors: 'In front of the Holy Cross parish school, the pastor, Father Brouchard, seized Nao Leng's hand: "Welcome! We've waited so long!"' (Garrett 1974). Since the days of Samuel Pollard, he might have added.

It has generally been the case in Asia that Christian missionaries have found it difficult to gain converts among the adherents of highly organized state religions such as Confucianism, Daoism, and Buddhism, and have consequently been forced to turn their attentions to smaller, more deprived minorities who have been grateful for the material benefits offered them. Indeed, in the latter respect, their achievements have often been laudable. But one must ask, I think, at what price, in terms of the divisions introduced both into minority communities and between them and others, such achievements have been gained.

Missionary Christianity: The Current Situation

In Thailand in recent years, the situation in which Christian beliefs are adopted in times of extreme economic deprivation, and the tendency for such adoption to be associated with the appearance of messianic ideals, has continued (see M. Heimbach 1976; Kuhn 1956; Scheuzger 1966). Cooper (1984) has observed how emergent poor strata of Hmong society would often adopt Christianity as a means of relieving themselves of expensive kinship or ritual obligations, such as the payment of bridewealth, and noted how Hmong Christians in Khun Sa expected a supreme force to intervene in their favour and were attracted by the 'saviour aspects' of Christianity. The custom of whole villages adopting Christianity together, often led by a senior elder, has also continued, and in the following chapter we examine more closely a case-study of the mechanics of this. The Catholic missions, who maintain a special school in Chiangmai for Hmong children, have in general been far more successful in winning adherents among the Hmong than have the many contending Protestant divisions, although some of the latter have been working in Northern Thailand since the 1920s. I believe one of the reasons for this is because the Catholics in Thailand, as one senior priest told me, have pursued a strategy of working with rather than against the beliefs of the people they are seeking to convert. This gradualist approach has fared better than the culturally radical techniques of many Protestant missionaries in Thailand, who have often encouraged the burning of altars and shamanic equipment.

It is partly also this contrast between the sympathetic and more conservative approach of Catholic missionaries (who have produced monumental transcriptions and translations of many Hmong rituals and some folktales) towards Hmong custom and belief, and the radical intolerance of the Protestant missionaries, which explains why messianic forms of Christianity among the Hmong have in general resulted from the encounter with Protestant rather than Catholic creeds. However, this in turn arises from a more fundamental distinction between the two approaches, in which Catholic missionaries interviewed betrayed an essential *scepticism* towards the bases of Hmong beliefs, which *allowed* them to adopt a more sympathetic attitude towards them, while Protestant missionaries demonstrated a surprising *credulity* about what they described as the 'demon-worship' of the Hmong, which similarly explained their more *intolerant* approach. Thus one Protestant missionary who had worked with the Hmong for many years asked me why I was wearing a 'devil's string' around my wrist, which was, in fact, the string from a protective ritual I had attended. Seeing my surprise at the expression she had used, she then questioned me intently as to whether I believed in the spirits worshipped by the hillpeople. When

I told her that I tried to keep an open mind on the matter, she assured me that she and all the missionaries she worked with did believe in these spirits, since 'otherwise we would have no basis to talk to their worshippers on'. There is a fundamental difference of attitude and approach towards Christian conversion between Protestants and Catholics, which in my opinion explains not only the greater success of Catholicism among the Hmong, but also the association of Hmong messianism with Protestant, rather than Catholic, forms of Christianity.

Hmong informants, for their part, confirmed this by broadly classifying Christian missionaries as Catholic, Protestant, or Seventh Day Adventist. With the Catholics, they said, one could smoke, drink, and perform almost all of the ancestral rituals, including the major funeral rite of *qhuab ke*, the 'opening of the way' for the reincarnating soul of the deceased, except for the practice of offering food to the spirits at funerals, which entails a symbolic feeding of the corpse with rice, pork, and whiskey three times a day. However, this exception was felt keenly, and I was told that because of this, 'If I became Christian, I would pity my mothers and fathers, and my grandsons would be ignorant.' The Protestants also permitted one to eat pork and smoke, but one could not drink. With the Seventh Day Adventists, however, one could neither drink, smoke, nor eat pork, and there was consequently some confusion between the Seventh Day Adventists and the Muslim Chinese, who likewise did not eat pork.

While in general maintaining, therefore, a healthy attitude of pragmatism towards these Christian divisions, there can be no doubt that the Hmong were far more favourably disposed and sympathetic towards Christianity as a whole, than they were towards the missionary Buddhism examined in the first half of this chapter. Christian missionaries had built schools and hospitals for them, encouraged and sponsored them to learn and become educated, took a genuine interest in their culture (at least as regarded Catholic missionaries), and bothered to learn their language. They were neither averse to living in Hmong villages, nor to having Hmong live with them in town. This was in great contrast to Thai missionaries of all descriptions. Moreover, the Hmong tended to see the Christian missionaries as representing a higher, more superior power to that of the Thai state, which might be appealed to in times of need, and which was favourably disposed towards the Hmong. This was exemplified by the rumours of myself being the agent of a Western power commissioned to give aid to the Hmong noted in Chapter 2 above, while the place which was at first found for me in the village clearly derived from the position of the foreign missionary in Hmong society, for which a special term has been coined, usually applied to priests only; *txiv plig*, or father of the soul, in contrast to *txiv neeb*, or father of the familiar spirits of the shaman, applied to shamans. Originally, villagers would inadvertently refer to me as *txiv plig*. And the superiority of the power which foreigners were assumed to represent had, of course, been reinforced by the known reception of Hmong from Laos by foreign countries.

In terms of personal belief, although I interviewed a great number of Hmong about their Christianity, I found no one who had completely rejected his faith in ancestor-worship or shamanism, although I did come across one who whole-heartedly believed in Christianity while not fully abandoning his original beliefs. For the remainder of those, however, who properly understood the rather abstruse Christian teachings to do with the Trinity and Original Sin, they certainly made earnest attempts to believe and, as I was often told, 'half believed'.

Despite this, there remained serious conflicts between the original belief-system and Christianity, which the case-study following this chapter is designed to illustrate. The majority of these conflicts had developed over the non-payment or minimal payment of bridewealth by families which had converted to Protestant Christianity. Further cases of this have developed among the Hmong populations overseas, which are considered in the final chapter.

However, one of the most serious conflicts which had developed in the Catholic Hmong Centre in Chiangmai concerned the prohibition of opium smoking, which some Hmong working there had argued was part of customary behaviour when in the mountain villages, and so should not be prohibited on such occasions. Another case involved a Catholic missionary of many years' experience in a Hmong settlement who had championed the cause of the first wife of a Hmong religious teacher's brother against the second wife he wished to take. Many Hmong argued that polygyny was a husband's right, and that the missionary was in the wrong to have intervened in customary practice by championing the cause of the first wife. Another large settlement had been seriously divided along clan and residential spatial lines by the impact of, in this case Protestant, Christianity. This had originated from the death of a Christian convert. Neither the non-Christian members of his own clan, nor the Christian members of other clans, could perform the proper funeral rites for him or have him buried, since he had relinquished the ancestral spirits of his own clan through his adoption of Christianity, while the funeral practices of other clans were different from those of his own and could not be performed for him. This case became so serious that it had to be submitted to the provincial governor, and the divisions in the settlement remained.

The primary motivation for Christianity remained, as transpires further from the case-study in the following chapter, to achieve some form of social or economic advantage, whether through being relieved of an insupportable economic burden, becoming literate either in the romanized version of Hmong or in Thai, or through being in a position to be able to appeal over the heads of the Thai authorities, to some superior authority. The primacy of economic determinants in Christian conversion is shown not only by the many cases in which impoverished families or members of such families adopted Christianity in order to be sponsored educationally by missionaries, but also in that, as the Chiangmai questionnaire showed, the majority of Hmong boys studying in Chiangmai at missionary establishments as at other establishments, were the *younger* sons of families. It is still the case that eldest sons are needed to work in the fields, while what extra-cultural advantages there are to be had are usually made available to the younger sons, as in the case of the youngest son of the old man of the focal village.

At the same time, old prophets and sages circulated ribald tales about Christianity and foreigners, or attempted to include elements of Christian teaching with which they were familiar in their own folk cosmologies and legendary accounts of the past. Apart from the old saw that Westerners ate babies (which is supposed to have arisen from the sight of two missionaries eating tinned sausages, but which may be read as a more serious allusion to the abstraction by missionaries of Hmong children), a story also circulated that foreigners were closely related to the Hmong, having descended from the daughter of a Hmong *Huab Tais* who, having no husband, instead wedded a monkey, and was expelled by her father, who cast her out to sea in a boat. Another anecdote went as follows (Samoeng, June 1981):

Jesus and Yawg Saub were like two brothers who were at war with each other: two Kings. Saub defeated Jesus in the end, but Jesus was able to fly up to heaven carrying a great wooden cross on his back. When Saub saw this, he realised what a great King Jesus must have been. As he flew up to heaven, a horse laughed at Jesus, who cursed it (*foom*) to be sterile. Then a cock laughed at him, so he cursed it too. And this is why ducks and donkeys cannot bear their own children.[15]

Saub is the *deus otiosis* of Hmong cosmology, often depicted as a kindly old man who appears at moments of crisis, such as before the Flood, to give aid and counsel to the Hmong. Sometimes attempts are made to identify Yawg Saub with Jesus (pronounced as 'Yesu'), since they sound somewhat similar, and these attempts have been fostered by missionary translations of 'God' as 'Saub'. Other attempts at identification are also made, as in the following extract from a longer account of creation and cosmology (Nyiaj Tub Yaj, Loei, March 1982):

We Hmong say that Jesus is Siv Yis. And when two thousand years are over, Jesus will come down to earth again. At that time, he will not care whatever people may believe, but will receive only the good people. When Siv Yis returns he will rule the earth, and everybody will be important. There will be no more war. When Siv Yis was here, he had a flying horse. . . .

Siv Yis is the premier shaman, the 'foundation' of shamanism, who is believed to return to earth one day. In such accounts, we see not only an example of the kind of poetic recombination examined in Chapter 3. While superficially trivial, they, in fact, represent serious conflicts of belief, and the attempt to express changing social circumstances in traditional terms. Less serious reactions to Christianity included a pun on a Thai word which sounds like 'Christian' but, in fact, means 'counting money' (*kitsetang*).

From the survey site, in addition to Xeev Hwm's youngest son in the focal village, the youngest son of Vam Xauv, the headman of Hapo, had also been sent to study in Chiangmai under Christian auspices. It was the latter who confessed to me that he felt ashamed of being Hmong when he was in town. There were also other isolated cases which contrasted with the communal adoption of Christianity described in the next chapter. One woman settled in the site had originated from Laos. All her agnatic relatives had been resettled in French Guyana, and occasionally (not being able to write) she would receive cassette tapes from them. In her youth her family had been cared for and partly educated by a famous French missionary, whom she remembered with great sympathy and respect, and she partially regarded herself as Christian as a result. Xauv, the case described in Chapter 2, who suffered extremely bad health and nightmares, claimed to have become Christian after he had prayed to the Christian God in his dream, and his illnesses had improved after his other prayers had been of no avail. But he did nothing about it, and as a practising shaman continued as before. I was once told, watching a shamanistic seance, that if I had been Christian I should not have been allowed to watch by the shaman who was conducting the session. In one of the outlying settlements lived a family whose head had been converted by the Seventh Day Adventists after his wife's inability to bear a child had been cured. Repeated miscarriages are thought of as the return of a single, playful child-spirit, and his wife's 'having had the same child several times', as it was phrased, was still attributed to unredeemed debts from a previous life despite the family's conversion to Christianity. These are the sort of isolated and perhaps not very significant cases which occur. What is more significant, in terms of forming a formidable alternative to the traditional bases of Hmong ethnicity, is where whole villages, or segments of villages, convert, and this, as we see in the following chapter, is by far the more common pattern.

In the previous chapters, we have examined three dilemmas, or sets of alternatives, which confront the Hmong at the level of the village. These are whether or not to cultivate the opium poppy, whether or not to support the Thai government, and whether to adopt Thai Buddhism or foreign Christianity. Obviously these are not oppositions of the same logical order, although they do represent the real emic alternatives which are likely to confront a villager. Underlying these sets of alternatives, however, is a deeper and more analytical opposition, between a fourth set of alternatives which we shall express as whether 'to be Hmong' or whether 'to become Thai' as Shan were said to have 'become' Kachin (Leach 1954, p. 9), since when we come to consider the ways in which the Hmong have distinguished themselves from 'other peoples' in the past, in defining their own ethnicity through a series of negatives, we shall find that such elemental distinctions always take place in a *locus* with regard to the *Chinese*, rather than the Thai state, but this does not matter, since it only reaffirms the primal opposition between 'being Hmong' and 'being other than Hmong', while it further emphasizes what I have said about the village's retention of its *conceptual* autonomy in Chapter 1. The *underlying* dilemma, then, is whether 'to be Hmong' or 'not to be Hmong'. (See Chapter 6.)

However, with regard to the *religious* alternatives between Buddhism and Christianity considered in this chapter, it is clear that Christianity has been adopted as a strategy precisely because it seems to offer a *way out* of this fundamental dilemma. While Buddhism has been largely rejected because it is so closely associated with Thai ethnicity, and the Thai state, Christianity appears to offer an alternative way of *remaining Hmong without being assimilated* by the state, through gaining an identity which is of a higher order and avoids the more mundane choice between continuation of ethnic minority status or total assimilation. As Kunstadter (1983) has recently aptly phrased it, this identity is one of the 'Christian hilltribe'.

Analytically, therefore, the sort of cases which were considered at the end of this chapter, where Christianity comes into conflict with elements of the original belief system and traditional practice, in fact represent a more serious opposition than the opposition between Christianity and Buddhism, particularly since the latter is, in any case, as we have seen, not a particularly viable alternative. And it is this encounter between Christianity and the Hmong belief-system which is examined in the following chapter.

1. 'Pantheism' is a more accurate and less dismissive term than the term 'animism' more usually employed.

2. Only three of the thirty-three were eldest sons of their families: the results of this questionnaire will be issued in a forthcoming publication.

3. From *wiseed*, magical (Thai); Burmese *wessah*; cognate with 'wizard'.

4. S. Pollard (1919) reports seeing a 'Romanish' priest on the road between Miao villages, and Savina's work (1930) is dedicated to a Father André Kircher, of the Society of Overseas Missions, who worked with the Miao of Yunnan (1863–1922).

5. Betts reported that S. R. Clarke had opened four schools in Miao and Chong-Kia areas in 1898 and 1899 (*Journal of the China Branch of the Royal Asiatic Society*, xxxiii, 1900–1).

6. This movement seems to have followed the usual practice adopted by the Hmong if new territories are sought or anything unusual occurs, of sending out scouting parties of two men from each village to investigate and report back. This is arranged by election and mutual consent in the usual egalitarian way. If any adult male Hmong summons the Hmong to war, everyone *must* follow his summons, whether they are of his clan or not. They may, however, do so slowly.

7. The name preferred by the Lolo, who are now known as Yi, is Norsu (or Nosu). See Dessaint (1980).

8. See also Hosie (1914, p. 167) on the anti-opium campaign.

9. The use of spoons, rather than chopsticks, has been one of the most consistent insignia of Hmong ethnic identity in recorded writings and historical accounts of them over the centuries, and still remains so in Thailand to a great extent. I follow Pollard in referring to his converts as 'Miao', although there is no doubt from the photographs in W. Pollard (1928) and elsewhere that they included Hmong.

10. See Keyes (1977, p. 55) on the Baptist conversions of the Karen in Burma.

11. It is significant that Pollard should use the word 'King' here, since this was probably the way it was phrased by the Hmong. We have already encountered similar formulations in Chapter 3 in the examination of motivation for political participation.

12. Cf. Latourette 1929 (p. 363).

13. *Rog Paj Cai*, the account of Yaj Txoov Tsawb, a nephew of one of the leaders of the movement, recorded and mimeoed by Y. Bertrais in Sayaboury (Laos) in 1972 (author's translation).

14. Hence, probably, the saying reported by Allton (1978) current at the time of the war: 'One eats flesh and not skin: one kills Han and not savages.'

15. The story refers to ducks commonly laying their eggs in hens' nests.

PART III

5

The 'Agon' of the Village

THIS pivotal chapter is intended to provide a focal field in terms of which some of the alternatives examined in the previous three chapters are drawn out and illustrated in the light of the more profound themes which occupy the remaining chapters. What follows is a type of 'social drama' (Turner 1974, p. 32) which I found also involved my own position in the village, and which I illustrate as far as possible through frequent excursions into minor scenes which illuminated the motives and actions of the main protagonists.

I will first describe the composition of one household in the village in detail, because in so doing we may clarify the main issues which divided the village. This was the household headed by Suav Yeeb (Figure 5), the ritual head of the Vaj clan, an old man in his seventies with a beard which reached to his waist, who diligently tended the three green parrots he kept on perches below the thatched eaves of his house, and of which the Hmong are so fond. It will be recollected that power in the village rested on a triumvirate formed by this man's oldest son, the headman, and the old man of the village whose daughter the headman had married, and from whom the main kinship links in the village emanated. The ritual Vaj head, however, described as being in the first generation, in relation to the headman who was in the second, was too old to take an active part in village affairs, which he left largely to his sons, although he was always consulted on important matters, and was in effective retirement. However, as a ritual expert his services were much in demand, and he might be called upon by members of his descent group as far away as the provinces of Tak or Nan if there was an important ritual to be conducted, or the exact procedure of a ceremony was in doubt.

It was he who, at the New Year ceremony in the village, had pronounced the ritual blessing on all members of the lineage as we circulated three times beneath the cord attached to a green bamboo sapling. Suav Yeeb maintained the largest household in the village, with twenty-five people, or more than a tenth of the total village population, but it was, as we shall see, a household of tragedy. They had moved into the village ten years previously from another where he had enjoyed a position of some importance, after a series of illnesses and misfortunes which, however, only continued in the new location. His age, and the fact that his eldest son had only been a teenager when he had moved into the village, twelve years

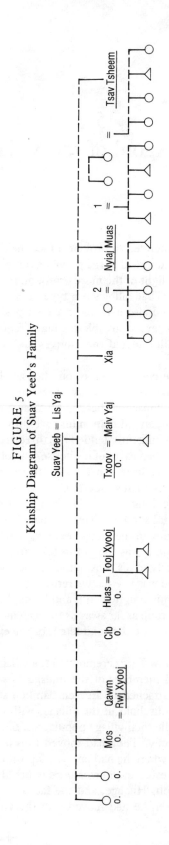

FIGURE 5
Kinship Diagram of Suav Yeeb's Family

after the headman, who had himself married matrilocally into the village, accounted for his ambiguous status in the village leadership.

Both in the focal village and in the larger settlement, illness formed at least as important a reason for village relocation as looking for new fields to cultivate. Often, of course, the two motives went together, and a general situation of not making enough to eat and being ill was blamed for relocation. But very often a resident may change village specifically because of prolonged or recurrent illness, or accident, since not only is it believed that the actual topography of the landscape in relation to house or village siting may affect or determine its inhabitants' prosperity, but also it is felt that by moving away one may elude evil or malevolent *dab* which may be associated with such a site. Thus, the entire outlying settlement of old Nomya, within the survey site, had migrated from its previous location after a series of miscarriages and stillbirths had been attributed to malevolent *dab qus* (forest spirits) living in a stream nearby.[1] Here we see that illness may form the primary reason for such migrations, even where the land is perfectly cultivable, and I believe this factor has been underestimated by most commentators on the shifting cultivation patterns of the Hmong.

At the time he had moved, Suav Yeeb had taken a new name for the same reason (his previous name had been Yaj Pov), so that the *dab* which had been troubling him would think he was someone else and go away. This, too, is very common practice to avoid calamity and misfortune, and certainly produces, in the material bureaucracy and among the takers of censuses, the effects it is designed to produce in the spiritual bureaucracy. It is similar to the practice of giving infants false or misleading names (and similar, too, to the Thai use of nicknames for the same purpose). There is much in a name, and as insignia of personal identity, like places, it is entirely appropriate that they should be changed in order to bring about or signify a change in the material circumstances attaching to a particular identity.[2] I have talked to informants who had had as many as four different names, and under extreme circumstances names of clans, too, may be changed.

Like other Vaj in the village, Suav Yeeb's family had tended to intermarry with the Yaj clan, although Yaj of a different area to that of the village or survey site. Suav Yeeb himself had married a Yaj girl named Lis, with a brideprice of nine silver bars and four Indian rupees, and their two eldest sons, Tsav Tsheem and Nyiaj Muas, had married (actual) Yaj sisters who were very nearly related to their husbands' mother, since the two brothers referred to their wives' father as *dab laug*, or mother's brother, although this was only in a classificatory sense.

Nyiaj Muas, however, had divorced his Yaj wife, claiming she was with child by another man, and had taken another wife from the Thoj clan into the household, with whom he had five children. His first wife continued to live in the village the family had moved from with two sons and two daughters. It is not wholly true to say for the White Hmong, as Geddes (1976, p. 59) does for the Green Hmong, that such children must take their mother's *xeem* (clan). Since there are strong imperatives for widows or divorcees to remarry, most Hmong would prefer to say that such children take their mother's husband's *xeem*, and this is, in fact, what had occurred in this as in other cases. Although two brothers marrying two sisters is an approved form of marriage, it is considered likely to cause trouble. The case of Nyiaj Muas' first wife was seen by many villagers as an example not only of the difficulties of maintaining two wives in the same household of the same generation, where they are not married to the same man, but also of the difficulties of two sisters marrying into the same household, since they may retain stronger attachments to each other and therefore the members of their birth *xeem* than to the families they have married.

More trouble had followed with the Yaj clan, moreover, for Suav Yeeb's youngest son, Txoov, had 'played' with yet another Yaj girl, who became pregnant, at an age too young for his father seriously to consider allowing him to marry her. Fathers have a large say in their sons' marriages, since it usually falls upon them to pay the bridewealth. Although a substantial fine was paid to the offended girl's family, as is customary in cases of pregnancy where marriage does not ensue, the fine was not considered sufficient by the girl's classificatory 'brothers' (*kwvtij*), and in revenge one of them had sexually assaulted Huas, Suav Yeeb's eldest daughter, who was likewise made pregnant. It was the incident of Txoov's unresolved affair with the young Yaj girl which, together with incessant agues and fevers among the rest of the family, led to Txoov, the youngest son, who could most easily be spared from the cultivation cycle, being despatched by his father to Chiangmai to be educated by Catholic missionaries, and the family's move into the focal village.

Misfortune was to continue, however. Suav Yeeb's wife suffered a severe burning while sitting in wet clothes by a fire near the rice-fields she had been working, and was more or less bedridden, at least confined to the house, for the rest of her life. Another daughter, Cib, committed suicide by swallowing opium after a serious row with her elder brother, Tsav Tsheem, who in a rage had torn her apron and accused her of sexual misdemeanour. This was said to have emerged out of his wife's not being able to find the razor to cut her hair, and accusing Cib of having hidden it.

Huas, with one son by the Yaj who had raped her, had married a Xyooj resident of the village who already had a first wife; that is, she had not formally married him, but had gone to live with him after becoming pregnant by him, which is permitted although not approved. She gave birth to two sons by him, one of whom was very fair-haired and fair-complexioned, with almost Caucasian features. This is a genetic oddity among the Hmong, which has been noted throughout the present century. Such children are sometimes said to have been *foom*, marked or cursed, and not to be trustworthy. I was sometimes mistaken (mostly by old women) for such a Hmong.

In fact, the clan or lineage system, the system of kinship, is the most important conceptual framework for a Hmong, and it is difficult to visualize social existence except in its terms. Even as a recognized foreigner, when I arrived in the village, I was frequently asked my *xeem*, and if at that time I said 'Tapp', it would be repeated with some difficulty, since there are no final consonants in Hmong, and become identified with 'Thoj', which is a Hmong clan name. This is a most important point, which is brought out in later chapters, since it implies not only that foreigners are expected to have clan affiliations, but also that those clan affiliations should be the same as the ones used by the Hmong. There is something very eclectic about the Hmong clan system in this sense, which may be characteristic of tribal formations in general.

The eyes of one family I visited in a neighbouring village lit up with immediate friendship and acceptance when they thought for a moment I had said I was a *kwvtij* (clan relative) of theirs, before I explained that I was *staying* with a *kwvtij* of theirs. It was not until I was eventually received into a clan, through a ritual involving the sacrifice of a pig and the tying of wrists with hempen thread, that I properly had a place in the village to which villagers could relate. If, as at first I did, I insisted on my original surname, I would then be asked how many people there were in my *xeem*, and if I then said there were about twenty, I immediately became an object of pity.

Huas would still help her original family, as well as her new husband's family, in the fields, and one day while clearing the fields, she accidentally shot her younger sister, Mos, who died. This is the story as I have it, although I suspect that the attribution of this death to Huas may be apocryphal, and another member of the family may have been responsible, since the gun went off as it was resting against a tree which was being felled. For shortly afterwards, Huas herself contracted typhoid fever and, after an extended and extremely expensive period in a Chiangmai hospital, partly funded by the missionaries, died.

Typhoid was quite common in the survey site: the little boy in the house in which I lived in the village had contracted typhoid shortly before I arrived, and I was given his sick-bed while he struggled for life in Chiangmai. Most fortunately he recovered.

Accidents in the fields of the above type are also quite common. An ongoing dispute, which was never satisfactorily resolved while I was in the village, involved one of the Yaj sons of the old man of the village, who had accidentally knee-capped a Vaj member from an outlying village, while in the fields. The Yaj claimed to have paid the equivalent of as much as 40,000 baht in compensation to the Vaj boy's family for the costs of hospitalization and transport to Chiangmai, and to replace the extra labour needed in the fields. The Vaj of that area, however, claimed the amount had not exceeded 10,000 baht, and the dispute was being arbitrated by the headman of the focal village (himself a Vaj). Oddly enough, the same man had shot another Vaj, in fact the mother of his sister's husband, in a similar accident three years before. Both, however, were quite clearly accidents, and the victim lay at home nursing his damaged leg, using the time to practise the *rab qeej*, the famous Hmong reed-pipe instrument (a distant relative of the Chinese *lusheng* and the Thai *khaen*), played particularly on occasions of death, in whose mournful yet appealing tones the entire repository of Hmong culture and folk-wisdom is said to be contained, and at which he would no doubt one day become a ritual expert.

After Huas' death, Suav Yeeb retrieved her three small sons from Tooj's establishment, which nobody could prevent him from doing since no bridewealth had been paid for them by Huas' husband, and in addition, it was said, she had not been properly dragged (*zij*) to her new husband's home, but had merely gone to live there of her own accord. Although a special form of wedding ceremony, *ua tshoob*, is related to *zij*, as we saw in the monk's account in Chapter 4, *zij* should form a part, if only a token one, of any *sib yuav*, or taking of a wife to the home, and, in a sense, constitutes a kind of protection for the bride, since it is easy for her husband's family to divorce her or send her away if it can be claimed that she simply moved in with them of her own accord, but much more difficult if it can be shown that the husband wanted her so much that he 'dragged' her to his home.

The payment of bridewealth, in fact, constitutes a similar sort of social insurance for the bride, since it cannot be demanded back in cases of divorce, abandonment, or other wrong committed against her. Moreover, it is very clearly distinguished from commercial transactions by the Hmong, who dismiss the very occasional outright payments of bridewealth in Thai currency instead of the traditional silver bars which have occurred as 'simply paying money and not even having a pig slaughtered, like buying an animal, not marrying a woman'. Thus the term for bridewealth is *nqe mis nqe hno*, or the money for the milk and food, tears, and labour expended by the girl's parents in bringing her up and washing her when she was a little girl. Moreover, strong links are maintained, where distance permits, between brides and the families of their birth. The daughter of the headman in the focal village, who had married into the larger settlement, used to visit her family not only as it is customary to do after the New Year, but also on other occasions, and would help herself to a rice-cake and throw some grain to the roosters as soon as she reached her first home. Bridewealth is a compensatory payment, then, which provides protection for the wife.

Huas' children thus retained their putative status as members of the Xyooj clan even after Suav Yeeb had taken them into his own household, because of a rumour that Tooj would eventually pay the bridewealth and retrieve them, although they did not like to visit their father, who lived at the other end of the village, because he would always take them along to work in the fields with him. In the end, however, Tooj took back the two youngest children in exchange for the oldest (not his own), who was about ten years old and was then given by his grandfather to his sister, who had married a Thoj in Nan province, to adopt, so that the dispute was resolved amicably.

In addition to his youngest son, who died in tragic circumstances I describe below, Suav Yeeb's wife had also had two children who had died respectively three days, and a few hours, after birth. The old man cried when we questioned him about his children. His older brother, who had settled far away in Sukhothai, was also now dying of old age. He could offer no real explanation of these misfortunes, although he was still sympathetic and grateful to the Christian missionaries who had helped his family. Some of the deaths among his children he blamed on *dab qus*, however, wild spirits disguised as animals, who prey on humans at night. Around the house grew tall clumps of banana trees and groves of bamboo, and from the ledge which skirted the front of their house, facing away from the village, one could see across the whole valley to where the diminutive Karen village of Mae Tien nestled, surrounded by verdant terraced padi fields, curls of blue smoke spiralling up into the mountains.

Because of these and other misfortunes, when Txoov returned after several years of education by the missionaries, including some time in pre-revolutionary Laos, he found the village at a crisis point in its identity. Txoov carried great status as the only literate member of the village (in both Hmong and Thai), and he began to teach his lineage brothers and the village children all he knew. Moreover, he had ideas about progress and development which he had formed during his years outside the village, through contact with Thai and Western functionaries, which he wished to see put into effect in the village. He was an imaginative and talented individual, who despite his Christian education had trained himself in all the procedures, riddles, and songs necessary to qualify as a wedding go-between.[3] His range of knowledge and experience was wider than that of the other villagers, and he was looked up to as a result. He was able to convince a majority of the villagers of the utility of his plans.

Three years previously, a French camera crew had visited the village for a period of three days. They had paid the villagers a large sum to stage a traditional boar hunt in the forest for them, and the spot where the boar had been killed was still remembered. They had then requested an elderly shaman, Xaiv Lwm, a Vaj, to allow them to attend a curing ritual which he was performing. This permission was granted. However, the crew had not made clear beforehand that they intended to film the proceedings. After the shaman's body had begun to shake and tremble as he entered a state of trance, studio lights were turned on and filming began. The shaman stopped abruptly, and removed the veil which covered his eyes, symbolizing, like Oedipus's blindness, interior sight.[4] He declared that his *dab neeb*,[5] or protective spirits, were extremely angry at this unjustified intrusion into their affairs, since they had not been informed first. On behalf of his tutelary spirits, the shaman demanded a large fine before agreeing to continue with the session, and a proportion of this was paid. Six months later the shaman died.

Although he had apparently died of natural causes, the villagers were unanimous in attributing his death to the wrath of his tutelary spirits, which had wreaked their revenge upon him. It was said that the French had been evil and 'like thieves' to behave as they had done, but also that the shaman had been stupid to agree to perform in front of them. For this reason, I was never able to take photographs of any shamanistic sessions in the village. Both before, but particularly after, the incident of the shaman's death, illnesses and fevers of every kind had plagued the villagers to an uncommon degree.

It was remarkable to what extent illness was *not* seen as an abnormal occurrence, to be restituted through appropriate rituals. Rather the 'way of sickness and death' was seen as part of the natural order of things, in a kind of fatalism which arose, however, directly out of real material circumstances. According to popular tradition among the White Hmong, one's wandering soul may leave one's body, its 'house', in cases of worry, fright, or shock, and fail to return, either because it is lost along the way or has been captured by malevolent spirits in an otherworld which is, at the same time, part of this world. This may also occur in the ordinary course of events, since the *tus plig* souls concerned are like children and like to play, as they do for example when their owners are asleep. Like children, they may wander too far from their houses and get lost, or fall down and be unable to arise again, or get into other kinds of trouble. In the absence of this soul, the body falls weak and ill, and a shaman must be summoned to recall it. In cases of family dissension, a shaman may diagnose that the *plig* of individual members of the family have wandered in different directions and, as at the New Year, must perform a wrist-binding ritual to bind the vagrant *plig* firmly to their owners and the household. If, however, the *plig* stays away too long, and the processes to recall it are ineffective, it may metamorphose into a winged insect and fly, ultimately, into another womb to be reborn.[6] In this case, as is only logical, its original body sickens to the point of death, and dies. In contrast with Hindu–Buddhist explanations of sickness and death, here sickness and death are seen as the *results* of reincarnation, rather than the other way about—death leading to reincarnation. This evidences graphically to what extent both are seen as part of an inevitable cycle arising from the natural order of things, rather than as aberrant breaches of nature or 'separations' in a Van Gennepian sense (Van Gennep 1960).

A similar reversal is apparent in the way possessive types of shamanism (*ua neeb muag dub*) are viewed, since it is always denied that such shamanism can be learned from a human agency, and attributed to the afflictions of the *neeb* (tutelary spirits) who cause an illness in the person they have selected to be a future shaman which, after other cures have proved ineffective, is diagnosed by a shaman as resulting from the *neeb* which have sought the patient out. The potential shaman must then seek a Master shaman, a *Xib Hwb*, who will teach him to control this experience. Proof that the *neeb* are, in Aristotelian terms, the *first* cause of shamanism, is deduced from the uncontrollability and lack of co-ordination in the first seizures of the new shaman, which I have witnessed, and in his clear and articulate enunciation during trance. If shamanism were learnt from *human* agencies, it is argued, one would expect older shamans to be much more word perfect than new ones, which is observably not the case. These *neeb* transmit themselves, although not necessarily in the direct line of patrilineal descent, along agnatic descent lines, together with the equipment of individual shamans.[7]

Every being is granted, before he is reborn, a 'licence' to live for a certain term, by Ntxwj Nyug, the malevolent Lord of the spiritual Otherworld. A shaman consulted may diagnose that his patient's licence is about to expire, but he can also enter into the Otherworld through trance and with the help of his familiar spirits and forces, beg, cajole, or bribe for the licence to be extended.[8] But the licence must expire at some point, and with it, the life of its owner. In a story told to illustrate the origins or 'roots' of shamanism, it is recounted how at the dawn of time Ntxwj Nyug was observed to be killing humans as fast as they were being born. Saub, the benevolent deity who has, however, to some extent become disinterested in the affairs of men, entrusts a proportion of his healing powers to a mortal named Siv Yis, who is thus enabled to cure illness and disease. Begged to return to earth after his ascent to heaven on his death, Siv Yis promised he would do so if called upon on the thirtieth day of the twelfth lunar month, but when the time came, realizing that everybody had overslept as he was half-way down the celestial ladder which joins the two worlds, Siv Yis cursed in a tantrum, threw his instruments of healing down on the earth, and returned to his heavenly abode. Some of these were picked up by different people, who used them, and so became shamans. Since Siv Yis promised to renew their efficacy if called upon at the New Year, this remains one of the primary spiritual rationales for the ceremonies conducted at the New Year, but the instruments were never so effective as when Siv Yis had used them, on account of this curse, and over the years their power has gradually diminished (cf. Mottin n.d.(b)). This account well illustrates the 'theory of deteriorating knowledge' among the Hmong which we examine below as part of their special view of history.[9] But it is also given as the reason why shamanism, nowadays, is not always, or only sometimes, as effective as it might be.

In the village this had become remarkably apparent. Illness and deaths continued with no apparent cessation, and the shamans themselves had become increasingly worried and angry at the inefficacy of their rituals. At the same time, Western medicines, and the hospitals in Chiangmai which many were now beginning to use, were recognizably effective.[10] More and more villagers were travelling to Chiangmai to purchase medicines of every description, and have themselves, their spouses, and their children, examined by doctors in the various clinics. One (Xyooj) villager purchased a pick-up truck to transport such passengers.

It needs to be emphasized here again that it was illness, together with the prospects of buying sacks of milled rice at government-guaranteed prices rather than the locally produced, Karen rice, which was of an inferior quality, which formed the *primary* means for contact between the village and the provincial centre. The village was by no means in the sort of position with regard to Chiangmai that some Hmong villages situated nearer Chiangmai were. Goods such as vegetables were only sold in Chiangmai on a one-off basis, usually to finance a trip there which had been undertaken because a member of the family was ill. The majority of villagers had no funds to visit Chiangmai, nor did they wish to, although they would be grateful for medicines brought back by other members of their clans.

The village did not participate in the handicrafts industry as many large settlements in other areas of Thailand did, nor did any tourists visit the village while I was there. Some Hmong villages have become partly or wholly dependent upon trade in Chiangmai, with trucks depositing villagers at Chiangmai's Night Bazaar every evening, and picking them up again in the small hours of the morning. This was by no means the case with the focal village, or indeed any of the villages in the survey site which, as I have said, retained their conceptual autonomy.

There is no doubt, however, that the impact of modern medicines did bring about a severe crisis within the village, for the villagers could simply not afford, following their customary life-style as in so many ways they were, to finance trips to Chiangmai by car and back, overnight stays in Chiangmai, or the prices of medicines and hospital attention. And shamanism did not seem to be as effective as the hospitals. One man, as reported above, became so disillusioned with his own powers as a shaman that he threw all his shamanic equipment away—gong and beater, mask and rattle, sword, bells, and bench—a culturally drastic move, since such equipment is usually inherited by other members of the lineage.

Into this situation, of economic and conceptual crisis, came the Christian Txoov, with his new ideas for change. As a Christian he wished the whole village to convert to Christianity, and he argued that if they did so they would no longer have to sacrifice their animals at shamanic curing rituals, while they could still continue to invest in the effective new medicines. Txoov managed to persuade his brothers and most of the rest of the village, including the Yaj group, to agree with him that they should convert. A few dissenters remained, such as the headman, who was jealous of the influence the newly arrived ritual head's family was gaining through his youngest son, but his wife was favourable to the idea and as a tolerant man he let it be known that he would not stand against the move. The Xyooj clan, likewise, were less favourable than either the more dominant Yaj or Vaj sections of the village, but would likewise have fallen in with the decision of the majority.

Txoov argued that since the villagers' hogs and fowls would no longer be required for shamanic sacrifice, they could be husbanded on a commercial basis for sale to Chiangmai, which slaughtered several hundreds of pigs daily for consumption. Thus the villagers would no longer have to depend on opium for a livelihood, which was in any case becoming increasingly more difficult owing to government restrictions and traders' credit demands. The animals would have to be penned, however, and this would entail moving the village to a flatter location about an hour's walk away, near the Karen settlement of Mae Sala, which was also on the main road which would facilitate transport. The village would also need to purchase a pick-up truck to transport fodder to the animals and the animals to Chiangmai, and a water-pipe to irrigate the site for the cultivation of other crops. The truck and the water-pipe could be purchased with the initial profits from the sale of the animals, and after that Txoov foresaw still further profits which could be used to invest in irrigated rice fields and a small rice-mill.

The plan was brilliant in conception and entirely coherent. It all depended, however, on the initial adoption of Christianity by the villagers. Moreover, his plans to relocate the village were opposed, and at least as regarded his own family, were vetoed by his two older brothers, who probably resented the position which their younger brother was gaining in the village, as well as by his mother, who wanted Txoov to get married and stay at home to look after his parents in their old age, as the youngest, favourite son should.

So Txoov did marry, another Yaj girl from the family's original village, thus to some extent healing the breach which had previously existed between the two clans there. But he continued to campaign for his ideas, and the village was divided over the issues involved, largely along lines of age, since most of the younger men supported his ideas for change, while it was mostly the older men who opposed particularly the proposal to relocate the village and to replace poppy cultivation by animal husbandry. It was at this critical juncture in the affairs of the village that, after a visit to Bangkok with a local missionary, and having with his new wife visited her parents on the way back, Txoov was murdered on his return to the village, just outside the village, by three Thai 'thieves' who took his watch, his silver neck collar, and the transistor radio he was carrying. It was, indeed, a crucial moment for the village, since after this time there was no more talk of Christian conversions. It was said that the Christian *dab* had not been strong enough to protect him against the influences of other, wild *dab*. At the time he was killed, Txoov had half completed writing a letter to the government about the water-pipe he proposed to construct to the new site he had visualized for the village.

The headman lost considerable face during this affair, since he was reluctant to go and report Txoov's death to the Thai police, as Hmong are often reluctant to do, and the matter was eventually reported by a close associate of Txoov who had previously lived in the village. The headman's reasons for not doing so were that he had not been properly requested to by the dead man's brothers or father, on account of the latent antagonism between the two families, although in fact he did go, secretly and of his own accord, several times to the police. Two of the thieves were eventually caught and arrested by the police, although the police accepted a bribe from one of them not to arrest them. The third, however, was never found, and the question of whether anyone had hired them to kill Txoov remained a mystery.

It was rumoured, for example, that he had had a meeting with local CPT cadres who had wished to recruit him shortly before his death, which had ended unsatisfactorily, and that they might have been responsible. There was certainly some local opposition to Txoov's plans, and like any charismatic leader, he had his enemies. He had left another son by a Lis girl he had failed to marry in a neighbouring village, and when I interviewed her father he actually said that Txoov deserved to die, because he had not married his daughter (although his own son, who bore the same name as Txoov, was in a precisely similar situation at the time with regard to another young girl). However, violence of the kind which killed Txoov is in such remote Hmong villages relatively frequent: the larger settlement maintained a permanent village guard throughout the year, and the local Royal Project officer never returned to his site after having been shot at by strangers in his jeep. It probably was violence of this kind which killed Txoov.

Txoov left a beautiful and charming widow of twenty-two years, with a posthumous son born a month after her husband's death, who was perhaps the most tragic member of this sad household. This child had had to be given two names, since it was born unexpectedly when Maiv, Txoov's widow, had returned to visit her own parents in her original village, where three days after its birth her father had had to arrange a soul-calling and naming ceremony (*hu plig*) for the child. Strong sanctions exist, however, against such matrilocal births. The child belongs to the Vaj clan, like its father, and should be born in a household belonging to its own clan, while here it had been born in a Yaj household, the house of its mother's original clan. So another *hu plig* ceremony had to be conducted for it after it had been brought to Suav Yeeb's household, and therefore the child was said to have two names.

The villagers continued to perform shamanic and other household rituals as if there had been no question of their conversion to Christianity. They would also buy medicines from Chiangmai and take the severely ill to hospital there whenever they could afford to. Sometimes they stated that hospitals were good for the body, while shamanism was good for the soul, although this may have been an older formula, since synthetic medicines have been to some extent assimilated to the category of herbal medicines, and it was also often said that while shamanism cures in the spiritual sphere, herbal medicine cures in the material sphere. However, I was also told most seriously that the best cures were those where one underwent a shamanic ceremony before going to hospital, and then another afterwards.

The villagers maintained an ambivalent attitude towards Christianity, which had seemed to come so near to solving all their problems, and a visiting foreign missionary was received with extraordinary deference, and housed and feasted. Many of the villagers knew a little bit about Christianity, and were not sure what attitude to adopt. The local missionary, who had hoped to see the entire village convert, was told by Txoov's two brothers that, while Christian sympathies remained in the village, the time was not now ripe to proceed with any further missionary work there, and the matter had better be laid to rest for some years. Yet even the headman of the village, who had not been particularly favourable to the introduction of Christianity, recalled the missionary teaching that the Hmong seemed to have been particularly singled out for suffering and persecution among all the peoples of the world. As for Txoov's father, he blamed his two eldest sons for not assenting to Txoov's plans, and wished they had agreed to move from the village when Txoov had wanted to. As the youngest son, Txoov had probably been his favourite child and he had laid aside a special store of silver for Txoov to inherit when he was older, which he said had now all been used up. He had not given it to his other sons.

My own position in the village was somewhat affected by these events, since on my original arrival I had been introduced to the headman, who had invited me to come and stay with him and live in his house as long as I liked. As he put it in a comment, the full significance of which I did not realize at the time, 'There have been many people coming here to try to change our customs, but you are the first who has ever come here who has wanted to learn anything about them.' My interest in 'custom' cast me as an unwitting conservative within Hmong society, but I failed to appreciate, at that time, to what extent the headman wished to offset the much more extensive contacts with the outside world which Txoov's family maintained, through his patronage of me. When a former associate of Txoov, closely associated with Txoov's family and a previous resident of the village, presented himself at the village, having heard that I needed a research assistant and I employed him, the headman told me privately that he could not allow him to remain in the village, since, (as I now see it) he was too closely associated with the neighbouring family for me to be seen as entirely the headman's protegé. I was eventually forced to move to an ex-Yunnanese shop-house in the neighbouring larger settlement, where I installed the assistant, with his family, and after that divided my time between the focal village (which I found easier to view objectively once I was no longer entirely identified with one family there), and the larger settlement, with which I became familiar in a less structured, and more informal way. I continued to spend the majority of my time in Nomya and found Hapo a convenient place to retire to without losing contact with the focal village.

The latent rivalry continued between the headman and Suav Yeeb's oldest son, who organized all the games played at the New Year and took part vociferously in village meetings, in a manner which contrasted with the headman's milder, more retiring manner. Towards the end of my stay in the village, however, the headman announced that he had had enough of the responsibilities of his position; he felt he was getting too old and weak, and felt tired or troubled at heart, and announced he would not apply for the official position to which a headman could be elected when the local Thai District Officer visited the village the following year (after postponing his visit twice because the roads were impassable). Tsav Tsheem, Suav Yeeb's oldest son, was, like the headman's own brother who had returned from the CPT, a broody, melancholic person, perhaps as a result of the many tragedies his family had suffered, but very sharp and said to be the only man in the village apart from the headman himself who was 'clever at solving problems on the mountainside'. This qualified him, ideally, for the position of village leader.

I hope that I have shown in the above, something of the interweaving of traditional practices and beliefs with modern and innovative ones which has led me to describe the village as a 'transitional' community, at the same time as illuminating some of the complexities of the alternatives which defined it. The method I have employed in order to do this has been through the analysis of one household in depth, since through multiplex relations the moral essence of the village had become concentrated in the lives of each of its many members.

It is clear that becoming a Christian, like joining the CPT, is, in general, a conscious strategy arising out of pragmatic concerns with the highest priorities of existence, in particular solutions to the problems of hunger, illness, and 'aliteracy' (cf. p. 124), thus avoiding the alternative of becoming assimilated by the Thai state which also offers solutions to such problems. Another genuine Christian, with whom I was very well acquainted, had been sent off as a child to study with the missionaries after no fewer than five of his siblings had died of smallpox or in infancy, and his parents had taken to opium smoking to relieve their grief, becoming more and more impoverished as they did so. In such cases, the adoption of Christianity is a desperate, or indeed the only, measure possible to adopt. One can see here how such situations of deprivation may also lead to strong messianic hopes.

At the same time, the faith in the efficacy of shamanism has proved strikingly persistent. Occasional or even chronic failures of shamanism may be explained with reference to the talents of particular shamans, or in terms of the Siv Yis legend recounted above, without seriously impairing fundamental cultural beliefs about illness and disease. The fact that Txoov's death could be explained in terms of a conflict between Christian and other *dab*, an integral element of the indigenous belief structure, by villagers who had suffered considerable exposure to Christian teachings both through the visits of missionaries and Txoov's own presence in the village, demonstrates the persistence of the essential bases of belief. In a similar way, with the Hmong refugees in the United States and France, one finds that where ancestral practices have been surrendered, this is often explained by the lack of house-posts in their new homes and hearths with which the *dab neeb* are associated, rather than admissions of loss of faith being made.

While the high values attached to health and literacy examined in this and the previous chapter remain, therefore, the most important incentives for the Hmong to become assimilated to the Thai state, Christianity has offered, and continues to offer, a solution which seems to offer these benefits without robbing them of their ethnic identity as 'Hmong'.

1. See the case of Tsav Pheej, Chapter 2; also Chapter 8.
2. Cf. Parkin (1980).
3. See Bertrais (1978).
4. Hence possessive shamanism is known as *ua neeb muag dub*, black-faced shamanism, in contrast with *ua neeb muag dawb*, shamanism which does not involve possession. There is also a third kind of shamanism known as *ua txheeb*, or a method of divining through drawing sticks, which seems to be of great antiquity.
5. The most general category of ancestral and other spirits, but here referring to the particular forces employed by the shaman.
6. There is a dependent cycle involving the three *plig* which wander, remain at the grave, or are reincarnated, which according to the shamans inhabit the body, which was likened to the three members of a household, each one leaving after the other because it did not like to be left, until at last there are none left, and the house falls into decay.
7. The practice of *ua neeb* (shamanism) is designed to help *others*. It is distinguished from the practice of *ua dab*, rituals performed by each householder at his own altar or elsewhere for *himself* and his family.
8. It is significant in this context that the shaman should often use Chinese or other words to address his supporting agencies, the *neeb*. I was told by ordinary villagers that this was because 'If I am Hmong and I speak another language, the spirits are afraid.' It was always denied very firmly in the survey site that the Chinese could or would practise shamanism, and seen as something which only Hmong could do.
9. Ohnuki-Tierney (1973) describes similar notions among the Ainu.
10. For example, novalquine was sold for headaches in the Chinese shop, and other forms of quinine used for other complaints.

6

The Importance of Literacy

WE have seen in the foregoing chapters, then, that at every level of the social formation—economic, political, and religious—the Hmong present is defined by a series of irreconcilable alternatives. It is out of these alternatives, and the various bonds and compromises between them which different Hmong households, communities, and villages in different areas may choose, that that agony of decision which produces the future, emerges. The future is thus revealed, not as a unilinear entity the course of which is predictable, but as a multilinear process, defined at every stage by a series of imponderable alternatives (see Friedman 1975). We may talk here of a series of possible or potential worlds, which define that world in which the Hmong live. The synchronic present, for the Hmong as for the rest of us, is thus, in a phenomenological sense, always 'coming into being'—that is, it has never ceased to exist.

For while we may posit that, in the 'real' world, the *primary* oppositions which determine the political and ideological zones are of an economic nature, since it is from an untenable or unviable situation as regards the actual business of getting 'enough to eat', and the attendant problems of illness and lack of education, that the alternatives of communism or Christianity present themselves (and the case of Txoov can be seen as yet another attempt to comply with government directives against the cultivation of opium poppy), nevertheless it is possible to subsume all the alternatives we have examined under a more general, more abstract one, which is, the possibility of 'being Hmong' or 'not being Hmong'.

This question, although arising directly from field data, with reference to the regional sub-debate on people 'becoming' other kinds of people (Leach 1954; Lehman 1967a, 1979; Moerman 1965, 1968a; Naroll 1968), is the corollary to that examined by Lemoine in his study of the Green Hmong (*comment peut-on n'être pas Hmong?* Lemoine 1972a, p. 22),[1] and that by Lombard-Salmon in her study of the Miao of eighteenth-century China (*comment peut-on être Miao?* Lombard-Salmon 1972, p. 278). In the present study, it is the essential question, and is one of a fundamentally ethnic, or inter-ethnic nature. However, when we turn, as now we must, to a consideration of some aspects of Hmong *historicity*, we shall find that, just as neither the present nor the future of Hmong social existence can be defined totally in terms of a unilinear model, so, too, history and the past itself are composed of a series of alternatives and oppositions, through which it is that the Hmong express their own ethnicity and sense of cultural identity.

And so we must question an arbitrary view of history, which always insists on its own view of the past. We must accept the fact that history has no reality outside of our conception of it—an idealist point of view, but one which cannot be avoided if we admit that any history is made up of an arbitrary *selection* of temporally ordered sequences of events to which we have assigned a particular significance, out of the total sum of past events (and non-events) which have occurred (or failed to occur). This view may, moreover, provide a perspective which allows history to enter anthropology in a new way, and is made up, together with our reactions to it (and particularly in the context of an oral tradition) of a series of conflicting viewpoints which are vital to our understanding of the process of history as it is. We shall see, in the consideration of tales and legends collected in the village which follows, something of how history is constantly refashioned in the context of an oral tradition, as we have to some extent already seen in the story of Tsav Leej in Chapter 3.

The two major themes which have emerged from the preceding chapters are the importance of the acquisition of health and the importance of the acquisition of literacy, beyond immediate subsistence needs, in the Hmong community. Since illness has already been examined to a great extent in Chapter 5, and is reverted to in Chapter 10, it will be appropriate here to consider some of the constituent elements of how the lack of literacy is seen. The following account of how the Hmong left China (by Xeeb Thoj, Hapo, June 1981), is remarkable in the strong historical awareness it demonstrates, within the terms of a purely oral tradition, of *literacy* and the lack of literacy. This suggests that the culture had had long-standing contacts with the members of a literate tradition, and had evolved means of explanation for its own differences from such a tradition, out of such contacts. The *loss* of literacy is thus attributed by the Hmong to the persecution they suffered at the hands of the majority Han Chinese, which drove them out of China and into the lands of Thailand and Indo-China. Hence, loss of territory is associated with loss of writing, the latter itself equated with the inability to pursue further study and become officials which we have already encountered in previous chapters.

This is why we Hmong have no books. It was like this. Long, long ago, Hmong were the eldest sons. They went to the fields to make a living for themselves, but they did not, could not, study books. According to the elders, a long time ago, everybody moved, and crossed the great waters. The *Mab Suav* (Chinese and others) carried their books across on their heads, so that they would be able to learn letters.[2] But we Hmong were so afraid of our books getting wet that we could not do that, and we were hungry, so we ate them all up. That is the reason why now we can only be clever inside, in our hearts and only remember in our hearts, not in books. But before that, we had books of our own. That was in China, where I have heard the Hmong still have books (writing).

Once upon a time in China, there was no wood for making fires or houses, and only earth could be used (for houses). And they say that there was one named old Saub, who knew one hundred kinds of things (everything). He was there, and he lived in the fullness of time. When he lived he knew everything, a hundred things, even the language of the tiny insects. On the earth, what people would come, fierce or clever, he knew it all. He lived in the fullness of time, and he said, 'I will return. You just wait for me, and eventually I shall return to teach you.' So the Hmong just worked their fields so they could eat, knowing no other kind of work, as day by day passed by.

Once Hmong were working their fields in the lands of the Great Dynasty (China) by the banks of a great big river. They had one dog which they could not find for a whole year, and when it returned it had burrs all over its head. They knew that if it had burrs on its head, there must be some good land to live off, and so they followed its tracks for a long, long way. It was such a long way that they could not see their way, so they came back to make a torch to light the way. A bee hanging inside a beehive in a tree; it could sting, so they put it near the fire to make a long candle, forty armslengths (ten daj) long, so they lit the candle, and were able to see the way.

They followed the dog's prints deep into the country, always moving on, looking for good land to be able to live on for all time, many, many times, until they reached this land and settled here to live.

Some like myself, in my own generation, went to Laos, following the river downstream. Some went up into the mountains to live there, and of those who went to Laos, some were able to study books and become officials there. We crossed the River Mekong into this country. Of those encircling Chiangmai now, all crossed over that water there, looking for fields to live off, and moved up into the mountains to settle down. Of those around Chiangmai now, a hundred families of Green Hmong and White Hmong came. We grandfathers are the ones who went before to clear the path, and look after the Hmong who come over to this country.

Before that we lived in the Great Land for a long, a very long time. Several times kings arose among us, but the Chinese were so continually ferocious that they were unable to reign, and we Hmong were unable to live there, so we came into this country.

Here, the lack of literacy is closely associated with the loss of territory and the flight out of China, while the acquisition of literacy is again associated with the power to become officials. We also encounter for the first time the distinction between 'elder' and 'younger' brothers, applied to differentiate the Hmong from the membership of other societies, in particular the Chinese.

Although we shall find the metaphor of sibling rivalry of increasing importance in the expression of the rivalry between ethnic groups, in this context the association of primary importance is that of the *younger* brother with the acquisition of literacy (like the Chinese), and the *older* brother with work in the fields (like the Hmong). We have already seen in Chapter 4 to what extent this metaphor is drawn from actual behaviour, in that it is usually the older brothers who are required to work in the fields, and the younger brothers (like Txoov) who are sent off to study if the opportunity is available. When questioned, the above informant explained that formerly the Hmong had been in the position of older brothers to the Chinese, working in the fields while the Chinese had had to look up to them and obey them, but that because the Chinese had had the opportunity to study, they had become clever and cunning enough to chase and cheat the Hmong out of their lands, so that the Hmong were now in the position of younger brothers, and had to look up to and respect the Chinese. He did not make the further assumption, however, that any acquisition of literacy had resulted from this new status![3]

The most extraordinary feature of the story is that a society with an oral tradition should be so *aware* of its own lack of literacy, that it should attempt to account for it in the above way. Here, however, one does need to distinguish the sort of illiteracy which may be the result of class differentials (see Bernstein 1971), from what I have called aliteracy, in reference to cultures which have never possessed any form of writing. The Hmong, in fact, fall into an intermediate category, for although there are no historically verifiable grounds for assuming that they once did have a form of writing, or kings, it would be unwise to disregard the strong awareness of, and familiarity with, literacy and literate peoples, over a long period of time, which such stories demonstrate, and there is a sense, as I shall show, in which the story can be taken as true, or at least 'real', which is what I am beginning to mean by 'real' history.[4]

Indeed, it is stories of this kind which would tend to demonstrate that, as we maintained in Chapter 2, the Hmong *have* always maintained relations with the members of more powerful literate societies, and that their social system should therefore be seen as a 'sub-nuclear' one of the type described by Lehman (1967b) for the Kayah of Burma. The homogenous isolation whereby the Hmong are often distinguished from other ethnic minorities in Northern Thailand (Hinton 1969) is, therefore, more probably the result of quite recent conditions created by the centralist freezing of national boundaries under the impact of colonialist authorities (Leach 1961; Gesau 1983). In Chapter 3, we examined some of the extensive relations maintained in pre-colonial days between uplanders (although not in this case the Hmong) and the rulers of lowland states, and now we see that stories such as the one above demonstrate a strong historical awareness of majority populations and their societies. For the story of the loss of literacy, in its essential form, is not a new one. Hudspeth, whom we met in Chapter 4 as colleague and disciple of Pollard, the first Protestant missionary to work with the Hmong, gives the following account (Hudspeth 1937):

Before the Pollard script, books and a library were unknown. The great majority of these tribesmen had never handled even a sheet of writing paper or a pen. They had heard that once upon a time there were books: a tribal legend described how, long ago the Miao lived on the north side of the Yangtze River, but the conquering Chinese came and drove them from their land and homes. Coming to the river and possessing no boats they debated what should be done with the books and in the end they strapped them to their shoulders and swam across, but the water ran so swiftly and the river was so wide, that the books were washed away and fishes swallowed them.

This was the story. When the British and Foreign Bible Society sent the first Gospels and these were distributed the legend grew—the once upon a time lost books had been found, found in the White man's country, and they told the incomparable story that Jesus loved the Miao. Only the imagination can conceive what this meant to these hillmen, some of whom travelled for days to view the books.

Thus, the extraordinary mass movement to embrace Christianity which we examined in Chapter 4 did not arise solely from a degraded impoverished people clutching at foreign straws of succour, nor even from their desire to hear more of this strange 'King' who loved the Miao. The initial impetus arose out of the belief that the Hmong literacy, lost of old owing to the Chinese, was to be returned to them through the agency of the Christian foreigners. Chapter 4 has already shown how, to some extent, this became a self-fulfilling prophecy: the missionaries did invent scripts for languages which had none, and their converts did receive opportunities to study and become literate. In this the Hmong seem not to have been alone, for the 'myth of the lost writing' is shared by other minority groups of the region, and in at least one other case, provided the rationale for the initial adoption of Christianity. Thus, there is a legend among the Karen of Burma that at the dawn of time, when the Creator was dispensing books to the various peoples of the earth, the Karen overslept and missed out on the precious gift of literacy. In some versions, they are given their book, but it is consumed in the fires with which they burn their swidden fields. When Baptist missions first arrived in Burma, their 'Book' was interpreted by the Karen as their own, precious 'Golden Book' which a prophecy had foretold white-faced strangers would bring back to them (Keyes 1977, p. 55). Consequently, the Karen became Christians in great numbers.[5] The Kachin also have a myth that they devoured their own writing in hunger (Gilhodes 1909), as do the Akha (Gesau 1983), while Graham (1954, p. 129) mentions that the Miao 'legend of a lost book' was also found among the Ch'iang of West Sichuan. Many of the Karen settlements around the survey site were Christian, and had a particular feeling towards Christianity, and Westerners, as a result. One old Karen in whose house I stayed produced a document of British citizenship dated 1923, issued by the British Consul then resident in Chiangmai, and asked, while we ate bees, if this meant that he would be welcome to live in England. All this, I would suggest, is because such minority groups formed in the past, and to some extent still conceptually form, part of an *ideal Chinese political system* which was characterized by the very prominent position of literacy, to which, like the Shan and Kachin (Leach 1954) and Chin (Lehman 1963) they aspired, and were to some extent admitted. The colonialist forming of national frontiers, together with the advent of Christianity, which was almost invariably directed, in China, Burma, Laos, Thailand, and Tongkin, at minority populations, merely exacerbated the divisions which already existed between literate majority populations and minority groups. The conceptual distance between minority and majority population was *increased* by the adoption of Christianity by minorities, which assured them of a more favoured position with regard to the Europeans who controlled even those areas, such as Thailand, which they had not directly colonized.

For the sake of illuminating the constancy of the association between the main elements of the legend among the Hmong, particularly that between writing and the figure of a 'King', together with the range of variations individual informants may allow themselves, I give here a third version, collected by a more recent missionary in Northern Thailand (Scheuzger 1966, p. 92):

Why ever did those horses have to eat the books of our forefathers, many, many years ago? These Meo kings were the first there were in the whole great northern kingdom. Indeed in those days we had a land of our own. A Meo king ruled over us. We were the most powerful nation on earth. But the wicked Chinese were more cunning than we. They fell upon us in great hordes. They had better weapons than we had. We fought bitterly and courageously, but it was in vain. The Chinese knew no mercy. They murdered, enslaved and pillaged. We had to surrender. But not quite everyone gave in; whoever could escape did so. When the exhausted fugitives came to a wide river they rested, leaving their packs among the bushes. They were all overcome with sleep. When at last they woke up—O horror—the horses had eaten up the Meo books! Not a single one remained. Since then we have possessed neither books nor script. . . .

The raconteur then describes the passage of the Hmong out of China through a great 'Lake', and concludes by saying: 'Ever since that day we Meo have been sacrificing to the spirits and are pledged to them until our Fuatai comes again and destroys the spirits and all our enemies . . . "Until our Fuatai rises again and comes to help us" they all cried' (Scheuzger 1966, p. 92).

It seems as though my own informant (whose version I prefer since it allows him to conclude that the Hmong are at least now clever in their innards) may have tried to combine two versions of the story, in one of which the books are lost owing to the waters between China and Indo-China, in the other of which the books are eaten up, either by the Hmong themselves or their animals, since Geddes (1976, p. 20) has provided yet another brief version, from among the Green Hmong: 'At Pasamlien, I was told by some of the people that long ago, when they were still in China, they had a book like the Chinese. But one day it got cooked up and was eaten by them with their rice.'

In Scheuzger's highly coloured and somewhat Kiplingesque version, we nevertheless see the association between a Hmong form of writing and a Hmong king (a *Fuatai*, an alternative pronunciation for *Huab Tais*, or Emperor) found in my own informant's version, clearly established, as well as the association between the lost writing and lost territory, blamed on the Chinese, which we find in all three versions.

Here we may begin to see the importance of the manner in which the Hmong define their own ethnic identity by contrast with that of the Chinese, through a series of negatives: the *absence* of writing, the *absence* of rulers, the *absence* of land, or states. It is pointless to attempt to reconstruct historical *truths* or *falsities* on the basis of such stories, but through a structural examination of the essential constituents of the tale we have isolated, we may arrive at an understanding of historical *reality*, as the Hmong today feel and express it.

For myths of this type, projected into a distant past as they are, nonetheless express important realities which are present in a symbolic form. In effect, they articulate a capacity for 'alternity' to a current situation which is of a prospective type, and which Steiner (1975) has related to the use of the future and conditional tenses in European languages. Thus, it is not surprising that, as we shall see, these notions of literacy should be associated not only with ideals of royal leadership, but ideals of royal leadership which are *messianic*, and thus prospective to a large degree, since they importantly articulate aspirations of a *utopian* nature. A myth of the type 'Once we had a king, a country and a writing of our own, but owing to the Chinese we lost them' may be read as a *conditional statement*, of the type 'If it hadn't been for the Chinese, we *would have had* a king, a country and a writing of our own'.

Exactly as occurs with the implied negative in the subordinate clause of the English conditional, the real negative (absence of a king, absence of a country, absence of writing) has disappeared completely from the mythological statement. This is a society deprived both of a state formation and of literacy, and highly conscious of this deprivation. It is to this deprivation, which we have attempted to examine in Part II of this book, that the legend points, providing as it does so evidence of long-standing contacts with cultures which *were* precisely distinguished by those things which Hmong society did not possess. And thus Hmong ethnicity is defined, through a series of similar, *negative* oppositions.

It is almost impossible to underestimate the importance attached to the power of writing, and particularly to the possession of a Hmong script, by the Hmong. Moreover, the claim to have rediscovered the lost Hmong writing by the Hmong messianic leaders who have emerged during the present century has remained a constant one. The invention of the first Hmong script by Samuel Pollard was undoubtedly one of the most important reasons for Hmong to become converted to Christianity at that time, just as the Romanized Phonetic Alphabet (RPA) invented by Protestant missionaries Barney and Smalley, in collaboration with the Catholic pastor Bertrais, in 1953, and taught by Catholic teachers in Chiangmai, was undoubtedly one of the most pressing reasons for Hmong from outlying districts, such as Nomya, to be sent to Chiangmai to study. Yet this script was not the only one available to the Hmong.[6]

Protestant missionaries in Thailand, engaged for a number of years on transmitting the Bible into the authorized roman script (Pollard's had been hieroglyphic), were suddenly instructed by the Thai government to produce an equivalent in a Thai script for Hmong, and as a result had to alter their entire approach towards constructing such a script, although they made no apparent use of the Lao script for Hmong already in currency in Laos, which could have been used to good effect since the Indic-derived Thai and Lao scripts are essentially similar. The possession of writing has strongly political implications, as the legends we have considered in this chapter acknowledge, through linking writing with states, territory, and kings. Although it is not formally expressed, there is disapproval on the part of the Thai authorities towards extensive use of non-Thai scripts for the minority languages within Thailand, since the measure of autonomy which the possession of an autonomous script assures, is recognized. Thus RPA is discouraged at an official level, where Thai literacy is preferred, or the writing of minority languages in a Thai-derived script, and in the construction of such a script no reference was made to the already existing Lao script for Hmong, since it is used and was sponsored by the Pathet Lao.

The Hmong in the survey site clearly recognized the political implications of the discouragement of literacy in their own forms of writing, and moreover linked it with the story of their loss of writing attributed to the Chinese: in ordinary conversation it would be said that the Hmong had no writing because 'the Chinese hadn't liked it', and then it would be added, 'like the Thai'.

There is also a Chinese script in use for Hmong used in China, and the Pollard script has survived in China. But it was the romanized script which carried the most appeal for the majority of Hmong in Thailand, since it did not imply the necessity of integration into the Thai state as the use of a Thai script would have done, while RPA held out the possibility of becoming literate in their own language on their own terms, or at least on the terms of foreigners who were conceptually remote enough not to pose an immediate threat to a 'Hmong' identity. It is instructive to realize that one of the major problems of the educational projects in the village described in Chapter 2, which concentrated exclusively on literacy in Thai, was the high value accorded by the Hmong to literacy in their *own* language.

The political implications of the possession of a form of writing, and its association in the myths and legends of minority peoples with the state formations of majorities, point to truisms about the nature of writing which I shall not belabour here (cf. Goody 1986, p. 120). Even Derrida (1974, p. 106), who challenges the value of the distinction between peoples with and without writing in his examination of Levi-Strauss's 'A Writing Lesson' (Levi-Strauss 1955), admits that 'Rousseau and Levi-Strauss are not for a moment to be challenged when they relate the power of writing to the exercise of violence'. The relationships between writing, hierarchalization, and the historiographic recording of events, are revealed very obviously in his critique, despite his attack (p. 135) on the 'myth of a speech originally good'. But that the power of writing should have been so well recognized by the border peoples of the state which probably accorded the highest status to writing ever accorded in the history of civilization, is hardly surprising.

Some measure of the importance which the Hmong continue to attach to the possession of a form of writing which is uniquely Hmong, may be gained from a consideration of the conflicts which have emerged among the Hmong refugees in the United States over, precisely, the *romanized* script for Hmong. For although, in theory, the script should work equally well for both Green Hmong and White Hmong dialect groups, in practice the script was invented for White Hmong, and the major dictionaries composed for Hmong in European languages have until recently been for White Hmong. Some Green Hmong in the United States have consequently demanded that linguists should work out a separate system for Green Hmong, and in addition to this the romanized script is finding increasing disfavour among the Hmong abroad, since the values accorded to sounds in it are not always consistent with those in the European languages they now must learn.

This brings the possession of a uniquely *Hmong* form of writing into still greater demand, and one has, in fact, been invented.

In considering this, we shall be led towards the consideration of messianism which is contained in the following chapter, since it is from the claims of having rediscovered the lost Hmong writing that the leaders of messianic rebellions against the state among the Hmong have particularly derived the legitimacy of their authority. As Lemoine, who provides an account of the various scripts invented for the Hmong language, has noted (1972c):

Struck without doubt by the importance accorded to the written document by the Chinese administration, the Hmong dreamed of a writing fallen from Heaven which would be their own. . . . According to the messianic myth, a King had been born, or would be born, to gather the Hmong together and deliver them from the tutelage of other peoples. The King or his prophet did not fail to announce that he had had a revelation of a script. (Author's translation).

In this context, it is highly indicative of the centrality of writing and not-writing to definitions of Hmong ethnicity, that the new script for Hmong, which is taught secretly to children and adults alike in the refugee camp within Thailand, by specially robed priests in a temple of their own construction, should be associated with a messianic religious movement which has evolved particular rituals, carved wooden statuary of Hmong deities and culture heroes (which again do not normally exist), and a variety of entirely original musical instruments.

Txoov was teaching this script before he died, and it was said that adherents of this movement had attempted, with support from wealthy Thai benefactors, to gain support from the Hmong of Thailand through conducting a census of Hmong villages in Thailand and imposing taxes. The movement with which this form of writing is associated is a highly articulate one, which is examined briefly in Chapter 10. I was told, for example, by one of the priests of this syncretic religion, which combines elements of Christianity and Buddhism with customary Hmong practice, that they 'wished to have a religion like other people' since people with organized religions had states and countries of their own, as they wished the Hmong to have.

The invention, or discovery, of the script with which the movement is associated is attributed to a prophet of the Hmong *Huab Tais* named Yaj Soob Lwj, originally of Sam Neua province in Laos, who announced the imminent arrival of the Hmong Messiah for 15 September 1967 in the large military base of Long Tien where he was resident (Lemoine 1972c). Lemoine describes how some later said that Yaj Soob Lwj had overheard an argument between a toad (representing the Hmong) and a lizard (representing others) in which the lizard declared that if the people were really Hmong they should wear red and white strings around their wrists to distinguish themselves, so that the population of Long Tien began to do so. Yaj Soob Lwj, however, was imprisoned by the military authorities and eventually, I am told, assassinated, after announcing that the millions of caterpillars destroying the rice harvest were the soldiers of the future King (Lemoine 1972c). Despite the apparent failure of his prophecies (cf. Festinger 1956), the script he revealed is still in use, and much work has gone into improving it. Lemoine (1972c) provides an early description of this form of writing, and I include the full description of it which I received in Appendix 5.[7]

The close association between the emergence of messianic leaders and appeals to a uniquely Hmong form of writing can, then, hardly be in doubt. Paj Cai himself, the leader of the 1918 revolt in Laos referred to in Chapter 4, was said to have distributed to his followers magical squares of white cloth inscribed with strange characters which assured their invincibility in battle, in return for which a tax was levied on the families of his supporters (Allton 1978). Moreover, in the account of Paj Cai's revolt by Yaj Txoov Tsawb (1972) recorded by Bertrais, it is said that on Paj Cai's return from heaven where he first receives his divine instructions, he meets 'four madmen who knew how to write', to whom he transmits his message by means of writing a single character. On his death he bequeathed a mystical writing engraved on copper to his widow, to be opened a number of years later, and prophesied that prosperity would ensue for the Hmong provided a red rain did not fall from heaven at that time. It is this script which is believed to have been rediscovered in the form of the new script which I have described emerging in the context of the refugee camps.

1. Arising from the question put to a young student by an old farmer: 'Are you Hmong [or] not?' (Lemoine 1972a, p. 22).

2. *Mab-Suav* has the general sense of referring to people other than the Hmong, but also a more specific sense in which, as Graham (1937) shows, *Mab* was used to refer to the Lolo (from *mang*, brave or fierce) and *Suav* to the Chinese, from a derivative meaning 'play' or 'be idle', since they lived off lands which they did not have to work. In the sense in which it was used in the village, however, *Suav* invariably referred to the Chinese, and this is the sense in which I retain it. One needs to bear in mind the occasional ambiguity of the term, however, although normally it is highly specific, since as we see in Chapter 9, a similar ambiguity pertains to the Chinese term for the Hmong, 'Miao'.

3. See Lemoine (1972a, p. 193) on the association of the lack of genealogical depth imparted by absence of writing, with an egalitarian-type social organization.

4. For early reports of writing among the Miao, see Vial (1898, pp. 33–6).

5. For the association of myths of literacy loss with Karen *Buddhist* messianic movements, see Stern (1968).

6. Lemoine (1972c) has provided a full description of other scripts in use by the Hmong.

7. A rumour that the lost Hmong writing was concealed in the intricate patterns of the women's embroidery points to the importance of other forms of communication in an oral culture. Embroidery patterns, costume, and the music of the *qeej* (which can be transmuted into linguistic tones) all provide codes of communication whose meanings can be 'read' by an initiate. The importance of such methods of non-verbal communication in an oral tradition should not be underestimated.

1 Nomya village.

2 Grinding corn in the village.

3 Carrying wood.

4 Newly cleared swidden fields near the village.

5 Gathering medicinal herbs outside the village.

6 Inside the house.

7 A Hmong mother and her three-day-old child.

8 A taboo placed outside the house to prevent strangers entering after the birth of a child.

9 Resting during a name-calling ceremony.

10 Carrying water.

11 Dressing up for the
New Year.

12 Boys and girls playing catch.

13 A village school.

14 The author with a
village elder.

15 Sales in Chiangmai.

16 The father of the Hmong King, Ban Vinai.

7
The Role of Messianism

WE have already examined the 1918 revolt of the Hmong in Laos, inspired by the messianic leader Paj Cai, who is said to have bequeathed the rediscovered Hmong writing to the messianic movement which has emerged among the Hmong refugees from Laos, in the context of their resistance to the Pathet Lao. This movement is considered as part of the general refugee situation outlined in Chapter 10. In this chapter we turn to an analysis of the symbols and cultural beliefs underlying Hmong messianism, since these are shown to be central to the definition of Hmong ethnicity.

Both Paj Cai and the leaders of the current movement in the refugee camps look back to a culture hero named Tswb Tchoj, who is believed, like other culture heroes such as Saub and Siv Yis, to have promised to return to the earth one day. Tswb Tchoj became the Hmong *Huab Tais*, or Emperor, of China. His downfall is attributed, like the loss of writing considered in the previous chapter, to the persecution of the Chinese. The legend of Tswb Tchoj is taken to be the central paradigm of Hmong ethnicity, as it is defined through the use of metaphors of kinship, in opposition to that of the Chinese. It is for this reason that it is always resorted to in times of messianic stress. As Savina (1924) puts it; 'Inconsolable at this loss [of their country] they are always waiting for the great chief 'phoa thay' (*Huab Tais*) who must deliver them one day from the yoke of the foreigner.' It is all the more remarkable, therefore, that the terms in which the legend should be expressed, should be so clearly the terms of a geomantic system for the siting of ancestral graves which has always been seen as peculiarly *Chinese*.

In this account by Xeeb Thoj (Nomya, November 1981), the references to writing desks, books, and pens are particularly significant in the light of the notions about literacy examined in the previous chapter.

The Legend of Tswb Tchoj

Long ago, when the Hmong lived in *Tuam Tshoj Teb* (the lands of the Great Dynasty: China), a widow gave birth to a child, who in time grew up to become the *Huab Tais* (Emperor) of the Hmong. At that time there was a Chinese whose mother had just died, and the Chinese went off to survey the *loojmem* ('dragon veins') of the countryside, to find the best place for her grave. He came to a dry, dry lake, which a clever old Chinese sage told him was the most auspicious site. This ancient sage knew the ways of heaven, and he had prophesied that there was to be a great war in the land. The Chinese (who was prospecting the *loojmem*) took with him a wooden staff which he used to prospect the landscape, and when he found that both the dragon veins and the Earth King[1] dropped into this dry lake, he stuck his staff into it (to mark the spot). The sage told him to return the following morning to see whether a drop of dew had formed upon the tip of the staff (for this would be a sign of the auspiciousness of the place). But the little child (who lived nearby) had overheard the two Chinese talking together, and early the next morning, before the first cock's crow, he went to the dry lake himself. As the sun shone red in the morning sky the little boy could see the dew upon the staff quite clearly. He seized hold of the staff, and shook off the dew from it. Although the head of the stick was still wet, when the Chinese came along later, he could see no dew at all. The next morning the Hmong boy, who was only ten years old, came and shook the dew off the staff again. And on the third morning, he did the same again, so that the Chinese thought the burial site must be an unsuitable one for his mother, and went off to bury her body elsewhere in the mountains. So the Chinese were unable to become kings (*Vaj*) and rule over the land. But because they were so savage and proud they were always trying to trick the Hmong boy who became the Emperor, and so they were eventually able to kill him.

Motifs of trickery and deception characterize this genre of story, particularly with regard to relations between different ethnic groups, and between mortals and immortals. The child was an 'orphan' (like the messianic hero Paj Cai), who would be expected not to inherit his father's clan name, but that of his mother or mother's husband. Many of the images derive from the system of geomancy for the siting of houses and ancestral graves, which is considered more fully in the following chapter. The *loojmem*, or dragon veins, are natural lines of energy believed to course through the landscape, which determine the relative auspiciousness of a particular site. One's fortune, and the fortunes of one's descendants, depends on the correct burial of one's ancestors. The site concerned is the most auspicious site of all, which will assure the descendants of the ancestor buried there of birth in an Imperial dynasty. The staff referred to is one commonly used for divination by Chinese geomancers, and the dew may well be a symbol of fertility. Dry lakes which magically fill up figure prominently in Hmong folklore. The story continues:

Later another Hmong boy was born who became Emperor. However, he was unable to reign very long because of his impatience, and so only reigned for forty or fifty years. The Chinese were losing in the great wars which were ravaging the Great Land (China), and they came to beg the Hmong boy who had shaken the dew to help them. When he went to take the Chinese staff, he told his mother that in the dry lake there was a wonderful shop with a writing table and a hundred marvellous things inside the lake. At first his mother refused to believe him, but when she was persuaded to come and see, she found it exactly as he had said. He told his mother to sit down at the table, since it was a table of silver and gold, and after she sat there he wove[2] around it so that both mother and table were thrown outside into the weeds, and his mother vanished under the weeds. Then a voice spoke out of the heavens, announcing that because he had buried his mother alive, he would reign for seventy years. But he was not able to (at once), and while war raged throughout the land he only tilled his fields. (And so he became impatient.) 'Why have I not been able to become King?' he asked. 'I buried my mother alive,

and the heavens said that I should reign for seventy years. So the heavens lied.'
And he made a cross-bow, saying, 'Since the heavens lied, I will shoot an arrow
into them.' The arrow that he shot flew straight into the heavens, making all heaven
and earth shake, like a newly built house (with no roof). All the Chinese knew (of
him through this), and so they came to ask the Hmong boy to help them in the war
they were losing. He summoned all his soldiers, who were as many as the bamboos,
as many as the blinks of an eyelid, and they went to join the fray. Only once they
fought, and all the land was safe. But because he had been so impatient he, too,
was not able to reign very long before he died.

The notion of burying the mother alive is the first example of the images
of violence and destruction directed against the mother in which the story
abounds. That the Chinese should seek military help from the Hmong
may be a reference to 'true' historical conditions, in that Miao and other
aboriginal forces were often mustered by the Chinese against foreign
invaders over a long period, and as late as 1874 Miao troops were deployed
against the Japanese in Manchuria. Here again, we see the significance of
the shop, or market, as a place of metamorphosis, exchange, and trans-
formation at the boundaries of the two worlds where the grave is situated,
as we saw with the legend of Tsav Leej in Chapter 3. The image of the
writing desk is a further example of the way in which a Chinese bureaucracy
was reflected in the Chinese Otherworld (Feuchtwang 1975) and thence,
through a series of invented oppositions, in that of the Hmong. The shooting
of the cross-bow seems to derive from the Hmong creation story of the
shooting of the nine suns which Lemoine (1972c) identifies with the Chinese
legend of Chen Yi, the Excellent Archer. The Hmong hero shoots at the
nine suns which are burning up the earth and crops, but the last one hides,
and cannot be persuaded to reappear until it hears the familiar crow of the
cock, which is endowed forever with a red crest as the first rays of the
rising sun fall upon it. In the version I received of this tale, it was explained
that this was why the cock always precedes man, and acts as a psychopomp
which leads the soul of the dead on its journey towards rebirth. There was
a third incarnation of the Hmong *Huab Tais*:

Much later, but still a very long time ago, a child was born named Tswb Tchoj,
and his elder brother was called Tswb Xyos Tuam.[3] Every day he would catch and
ride a huge *Xib Nyuj* (rhinoceros) which came to eat up the (growing) rice.[4] Again
the Chinese whose mother had died came to sight the dragon veins, and again he
came across the (most auspicious) place. But this time it was guarded by a great
rhinoceros, like a bull, so huge and fierce that nobody dared to approach it. The
people who lived nearby that place told the Chinese that nobody could catch it,
except for one, very poor, orphaned boy. Anybody else who came near it would be
tossed into the marshes. After watching how the Hmong boy rode the rhinoceros,
the Chinese went home for his mother's funeral. He burnt his mother's bones to
ash, and made a powder of the ashes which he mixed into a flour-cake.

The Hmong boy went and asked his mother who his father was, but she told him
that he had no father. 'Then how is it that I am here?' he asked. 'Tell me truly,
mother, what person or animal begat me, or I shall take your own bones and burn
them.'

The mother, frightened to hear her son talk like this, told him that once she had gone to sleep near a boar which had breathed so heavily on her that she had conceived. She told him that she had buried the boar in the ditch around the house. So the child dug up the bones, and out of their ashes he made another flour-cake.

When the Chinese returned, he gave his own flour-cake to the Hmong boy, and asked him to offer it to the rhinoceros. But the boy offered it instead his own flour-cake, but the animal refused to open its mouth. It would only open its mouth when he offered it the Chinese flour-cake, because it contained the bones of a person, and not the bones of a pig.

So he offered it the Chinese cake, but when it opened its mouth, he threw in his own cake instead. However, the rhinoceros excreted it without it going into its stomach.[5] Twice he did this, and the third time he slapped the beast on the side so that his own cake went down into its stomach. Then he pierced the Chinese cake with his staff.[6]

Then the heavens and earth grew dark, as if the sky was about to tumble down. Darkness was everywhere, and an earthquake shook the land. And the Chinese announced that Tswb Xyos, the King, would rule over the Golden City for ten thousand years.[7]

But because the mother of the boy spoke up inopportunely at this point, exclaiming that her son was to be king, Tswb Xyos died.

The question-and-answer exchange between the mother and the son is a type of riddle which analytically represents the crux of the story, since it so clearly expresses the dubiety of paternal descent attributed to the Hmong orphan, and is repeated three times before the answer is given. Again, the images of violence and destruction directed at the mother recur. However, in the oral telling it is the threefold ordeal of offering the cake to the animal (whose acceptance of it would signify the readiness of the earth, in the site guarded by the beast, to accept the remains of the deceased ancestor contained in the cake) which really constitutes the climax of the tale, since it is always told with much repetitive gusto and physical enactment. Again, here, the themes of trickery and deception which characterize inter-ethnic relations emerge. The episode ends on a note of bathos marvellously characteristic of Hmong stories, as well as marking the third time images of violence and resentment are directed against the mother. Very often tragedies are attributed to the inopportune words, or curiosity, of women. If there is a European genre to which this sort of story can be compared, it is the tragedy.

The rhinoceros is now extinct in China, and neither this raconteur nor other informants could tell me exactly what a *Xib Nyuj* was, except that it was like a huge, fierce mythological cow. Although there is a Hmong term for rhinoceros (*twj kum*), the term used here derives from the Chinese. The rhinoceros seems to have always had something of an erotic significance, closely connected with water and fertility. The image of the small boy riding the rhinoceros is irresistibly reminiscent of representations of Lao-Tse, the inspirer of Daoism (sometimes said to have been a Miao tribesman) who is often depicted riding a water buffalo. These depictions of Lao-Tse may themselves have derived, however, from the Buddhist series of 'ox herding' illustrations, while the burning of the bones would also suggest some Buddhist influence. Hmong culture has been importantly influenced by Daoism, particularly through the medium of shamanism, and ancestors of the Hmong were almost certainly involved in the messianic attempts of revolutionary Daoist movements to overthrow the Qing dynasty and restore the Ming (cf. F.-L. Davis 1977). Miraculous births are not inappropriate in this Chinese context of messianic royalty, since many miraculous events were associated with the birth of Lao-Tse (Dore 1938) and other heroes. Analytically, however, the miraculous conception further emphasizes the doubts of paternal ancestry the story so importantly expresses (Leach 1966).

The story concludes:

He would have arisen again to fulfil the prophecy and truly reign, but the Chinese would not allow it. The Chinese placed a big piece of iron under his neck in the grave, but seven days later he was still rubbing his neck against the iron as if he would break it, so they went and put a big rock under his neck and reburied him in a 'chameleon' (a small) mountain. There he could no longer rub his neck, because he had rubbed all the flesh off it, and the flesh grew into the rock. So then the Hmong had no King, and only the King's younger brother was left. The Chinese killed the younger brother and burned his bones to ashes. Once more he was reborn, but the wicked Chinese had spied it out. They killed him and burned his body again and pulverized it into dust. They threw the dust into the mighty river, which flows all the way to America, so that our King has gone to be the King of America.

We could not stay in the Great Land because the Chinese were so ferocious and persecuted us. But Tswb Tchoj, the younger brother, was reborn as the King of America because his elder brother could not arise.[8]

Four times the Chinese persecuted the Hmong Emperor, who became a mighty bubble, floating down the river to America.

A very beautiful American maiden came down to the river to bathe, and the bubble burst against her body, and became a child within her womb. The girl was a virgin, and when her belly grew big, her parents scolded her and asked what had happened. When she explained, they said that in that case they should just wait and see what kind of child it might be. For three years they waited before the child was born, and it was born carrying a book and a pen under its arm, and became the King of America.[9] The child became the Emperor of America because the Chinese had been so fierce in the Great Country.

Iron is particularly inauspicious if placed in graves, since all that is buried should perish before reincarnation can occur, and nowadays not even synthetic materials should be used to dress the corpse for this reason. The opposition between stone and metal is examined in the following chapter, while, as we see later, rules about what should be buried in graves have posed particular problems for Hmong refugees in the United States.

This is a strikingly original ending to the story, which illustrates what I have said about the power of the oral tradition to blend disparate elements, and encapsulate changing social conditions. It is also important in that it provides a version of the myth which is as 'genuine' as any other (Levi-Strauss 1963b). The four consecutive episodes of the story function as variations on an essential theme of reincarnation which concerns the same cultural hero. Thus, we consider these as variant or parallel versions of the same structural elements. The ending, in fact, combines the most modern elements with some of the most ancient, such as the miraculous conception of the virgin in water. We are told of the most ancient peasant beliefs in China, that 'they imagined that virgins could become mothers by simple contact with the sacred waters' (Granet 1930). Within an oral tradition, such individual variations and departures form, in a sense, the substance of the tradition, and provide much of its strength and resilience. In the present case, the final incarnation may be seen as a reference to General Vaj Pov, the Hmong leader from Laos who is said to be 'like a King', now resident in America. There is a poetic logic in the employment of the Tswb Tchoj legend, which deals with the flight of the Hmong out of China, to include this latest and most traumatic exodus to the United States and elsewhere. On the other hand, 'America' may not have meant very much to the informant, except as a distant place to which he had heard that many Hmong were now going. But this is the stuff of which legends, like dreams, are made.

It must be emphasized that the tale may perform different functions in different contexts. It could be used as evidence that the Hmong King had now appeared and was summoning his followers to support him, as it was in Thailand during the early 1960s. It could be used to hold out the utopian promise that Tswb Tchoj would, one day, reappear. More often, it was simply told, like the story of the loss of literacy in the previous chapter, to explain why it was that the Hmong, today, had no King. According to one informant, the Hmong *Huab Tais* had already arisen: he had been born thirty years previously, but was now hiding in a cave because so many of his predecessors had been killed, waiting for a mighty war which would destroy most of the nations of the earth and be signalled by the appearance of a comet, before declaring himself. And, as we have just seen, the story may be told with a quite different conclusion.

This is because, I would maintain, the tale is open-ended in an important way, and might assume either an active, or a passive form. In the form in which I collected it, in the village, the tale was in a *latent*, or *passive* state. Its central importance to Hmong ethnicity, however, is shown in that it is always to this legend, and to the figure of Tswb Tchoj, that the leaders of the messianic movements among the Hmong appeal to sanction their claims to authority. When Paj Cai inspired a full-scale rebellion, claiming to be guided by the spirit of Tswb Tchoj, and when Yaj Soob Lwj announced the imminent birth of Tswb Tchoj, which his followers in the refugee camp today have symbolized by erecting a large statue of the black boar said to have been Tswb Tchoj's father, then, I would argue, the legend has taken on an *active, manifest* form.

The emergence of messianic movements in situations of encounter and confrontation between subsistence and market economies has been re-marked frequently enough for it not to need further elaboration here (Worsley 1968; Burridge 1971). What is remarkable in the present instance is the frequent recurrence of such movements throughout recent Hmong history, and the abundance of legends about Hmong kings, emperors, and messianic leaders, only a small proportion of which are considered here, among a people characterized by an egalitarian social organization and, apart from the role of the shaman, the lack of any significant social stratification within it.

Unfortunately, I cannot, for reasons of space, reproduce here all the variants of the tale I have access to. But the main outlines of the story remain very consistent. An orphaned Hmong boy seeks to outwit the Chinese in the choice of the ideal site for a deceased ancestor, which will assure him or his descendants of Imperial status. He is revealed to be the product of a miraculous union between a Hmong woman and a boar, and he is tricked out of his rightful inheritance by the Chinese. The manner in which the Chinese assure the downfall of the Hmong emperor, by burying him in an inauspicious way, refers to actual historical conditions, in that the graves of ancestors of pretenders to the throne or those severely disgraced were often desecrated in this way by the Imperial dynasties to ensure that the claims of their descendants could not be successful. Chinese graves were similarly desecrated during the Paj Cai revolt in Laos. In a valuable comment, the story has been described to me as importantly about 'themes of sovereignty and rebellion'.[10]

The most remarkable feature of this story of the Hmong orphan who becomes the Emperor, which accounts for the defeat and progressive marginalization of the Hmong at the hands of the Chinese over a period of time which is clearly seen as historical, is that the entire story should be couched in the idiom of a system for the geomantic siting of ancestral graves which has always been seen as a specifically Chinese one. Indeed, many of the images which do not communicate themselves at once may be explained in terms of this system. The essential principle of the system is that *it is upon the welfare of ancestors that the fortunes of their descendants depend*, and that this is importantly affected by the location of their burial sites according to the 'veins of the dragon', or currents of natural energy which run through the elevations and depressions of the landscape, particularly mountains and watercourses (Feuchtwang 1974; Freedman 1966, 1968; Durkheim and Mauss 1963). The system is known to the Chinese as *feng-shui* (winds and waters) and to the Hmong as *loojmem* (dragon veins). If good fortune is to be assured, the graves which are the dwelling places of the dead must be correctly aligned, as also must be the houses or villages which are the dwelling places of the living (Feuchtwang 1974, p. 17). Here, however, it is the dwelling places of the dead which particularly concern us.

In the legend of Tswb Tchoj, the contested burial site is the most auspicious site it is possible to find, which will assure the person who buries an ancestor in such a place or their descendants of Imperial status. Tswb Tchoj himself is, as we have seen, the figure whose advent is announced by the prophetic leaders of messianic revolts among the Hmong, who validate their claims to the sanction of Tswb Tchoj's authority through the proof of the possession of a Hmong form of writing. The entire story is about Hmong opposition to the Chinese. Thus, it is remarkable that the very means employed by the Hmong to justify their differences from and opposition to the Chinese, should fall so clearly into the pattern of a generic type of geomantic story common in southern China.

Thus, Graham (1955), considering a Lolo story about a great Lolo hero who was tricked out of his power by the Chinese, which validates the Lolo practice of cremation but otherwise bears no resemblance to the story of Tswb Tchoj, notes that such stories

... belong to a type quite common in China. The Chinese parallels can be called 'tales about geomancy'. Supernatural signs indicate that an emperor of a new dynasty will be born. Only one person knows the real meaning of these signs and the right place. He tries to bury his ancestor in the right place so that his child will be the ruler, or he tries to get the symbols hidden in the place for himself. The plan almost succeeds, but by some interference a catastrophe occurs in the last moment. It is surprising that these stories, which occur in almost all parts of China, have been also accepted by non-Chinese natives. ...

But it is even more surprising that such a genre[11] should have been the vehicle for the expression of historical *opposition* to the Chinese, and defeat by the Chinese, let alone have become the sanction for messianic uprisings against the state. The major hypothesis tested in the remainder of this study is that such legends, and the systems of burial and kinship with which they were associated, were not *adopted* from the Chinese, but rather formed part of a common system arising out of the interactions between many different ethnic groups in southern China. For the crucial import-ance, and centrality, of the figure of Tswb Tchoj, to Hmong culture and definitions of Hmong ethnicity, cannot be underestimated.

Thus, a ritual used in the village which may be performed in the case of the failure of crops, contains the following words:

Oh mother and father of heaven, greater and lesser emperors, great Emperor Tswb Tchoj and lesser Tswb Tchoj, Siv Yis (the premier shaman), Ntxoov Ntxooj (head of the shamanic spirits), Ancient Dragons, Lady Sun, Lord Moon, now there is sickness, now there is weakness, now there is complaint and disputation, I pray you to come, govern sickness and weakness, restrain gossip and idle rumour, by day or by night, I finish my plea.

Here Tswb Tchoj inhabits the pantheon of ordinary villagers, and is not reserved solely for situations of messianic crisis.

On the other hand, the prayer attributed to Paj Cai himself by Yaj (1972) runs as follows:

I pray to the heaven and earth, Imperial Mother and Father who reign in the City above, Lij Lim reigning throughout the world, now arise and take up your position, Great Dynasty of Heaven, Lesser Dynasty of Earth, Great Tswb Tshoj, Lesser Tswb Tshoj of the soldiers, now, rise up and regain your position, seated on the field of battle, to the depth of three cubits, I beg you, come down and help your sons and daughters, that they may gain their lands.

Here Tswb Tchoj is directly called upon to aid Paj Cai in his military struggle. According to the above account, Paj Cai was initially summoned up to heaven on a flying horse where, behind a great writing desk, he is informed of his mission by Tswb Tchoj and five other beings. Moreover, at the failure of his rebellion, Paj Cai attributes the reversals he has suffered to the deceptions practised on Tswb Tchoj by non-Hmong rulers, in a direct reference to the legend of Tswb Tchoj presented here. Thus it is, above all, to the figure of Tswb Tchoj that the leaders of messianic movements make appeal. Furthermore, both Tswb Tchoj and the historical figure of his prophet, Paj Cai, are associated together in the village. Thus, in the case of actual sickness within the household, villagers could *fiv yeem* (make a promise to heaven of future return) in terms which call on the Ancient Dragons and Venerable Xob, the Thunder God, the Lady of the Sun and Lord of the Moon, the Head of the Shaman's Forces, grouped together with the Greater and Lesser Tswb Tshoj, because they had 'all once been chiefs or officials'.

I would argue that it is *because* the legend of Tswb Tchoj so primarily expresses the cultural and ethnic opposition of the Hmong to others, that it is employed to articulate actual or current conflicts between the Hmong and members of other cultures, which consequently adopt a messianic form. The legend provides a paradigm of Hmong culture, which is activated at moments of extreme confrontation with other ethnic groups. While the legends I have presented were collected in the village in a *latent* form (although inter-ethnic conflict was already severe, as Chapters 2–4 of this book demonstrate), the legend has taken on a *manifest* form in the refugee camps, through the intensified interaction with, and opposition to, the members of more dominant ethnic groups during the war in Laos.

In the case of the conflict inspired by Paj Cai, the legend had also assumed an active form, since Paj Cai claimed to be guided by the spirit of Tswb Tchoj, and this was probably also the case with the revolt led by a 'Sioung'[12] Hmong King in 1860, whose actions remarkably resembled Paj Cai's (Lunet De La Jonquiére 1906):

There was at this time a man who called himself 'Sioung' of the Meo race, who became king, taking as title of reign the name 'Choen-Tien' and living on the mountain of Tang-Chang (the actual massif of Phia-Phuoc). Before becoming king, one saw him each day burning perfumes and sacrificing to the spirits. He used to practise the most prodigious leaps. Having piled up branches one on top of the other and made out of them a sort of very high tower, he would arrive by jumping in a single bound onto this tower from the top of which, seeming like a man who had lost his reason, he would cry, 'It is I who am the king of the country, the four spirits whom I command go to sew haricots and turn them into soldiers; nobody will be able to vanquish me.' All the officials went to Tang-Chang to see him and bring him presents. Sioung replaced the old construction of branches by an even more elevated tower of stone. All the inhabitants, seeing him rise from the earth and, with a single leap, jump to the very top of the tower, were seized with fear. They prostrated themselves at his feet and recognized him as their king. He began to erect a palace of gold surrounded by a wall of defence. His officers and soldiers all wore white turbans. Since the authority of Sioung was meanwhile not recognized by anyone other than the Meo, the Nung, and the Man,[13] he resolved to punish (castigate) the others. He went first to Lang Dan. . . . There he massacred a large number of Tho and pillaged all the houses. From there he proceeded to Quan Ba where he put all to iron and blood, such that one of his band intercepted some Tho fugitives on the Ha-Giang route and killed more than a thousand. The King Sioung returned to his palace, where he was worshipped while the rebellion continued . . . then, having killed his wife, the second daughter of one Tao-Yao, he wanted to marry the younger sister of the victim. The father-in-law opposing this fled to China and had the king killed by a ruse some time after.

This uprising should not be confused with those of the Hmong against the French which, according to Lam Tam (1972), broke out in 1904, 1911, 1917, 1925, 1936, and 1943, of which the most important were those of 1917, led by Paj Cai, and of 1911, also led by a shaman named Xyooj (cf. Yangdao and Larteguy 1979), which is described as follows (Lam Tam 1972):

This rebellion was carried out by the Meo in Dong Van district, Ha Giang province. Crushed under taxes, forced labour, and misery, they revolted against the colonial policy of monopolization in the commerce of salt and opium and restrictions in poppy cultivation. Under the lead of a poor local peasant, Sung Mi Chang, they demonstrated in front of the office of the French administrator, then launched an armed insurrection. With their flintlocks, crossbows and knives, the insurgents succeeded in capturing three out of four frontier posts. But their victory did not last long. At the beginning of the following year, the colonial troops counter-attacked. Sung Mi Chang was captured and his men scattered.

The account of the 'Sioung' rebellion above, however, shows remarkable similarities to that of Paj Cai later. Particular points of resemblance are in the construction of a tower, the great leaps practised by the leaders (shamans also practise leaps backwards into their benches), the sacrifices and burning of incense, and the attribution of the final defeat to a marital or sexual transgression. The failure of the revolt led by Paj Cai, who used to climb a seven-branched tree to receive his instructions from heaven, and his death (mourned by a tiger and a deer) was likewise attributed to the violation 'like dogs and pigs' of a group of Lao women who had surrendered to Paj Cai's General. Moreover, Paj Cai's death was prefigured in the flight of grasshoppers from a magical trunk, which irresistibly recalls Yaj Soob Lwj's announcement in 1967 that the caterpillars consuming the rice crops were the subjects of the future King (Lemoine 1972c).

In its active manifestations, then, the legend of Tswb Tchoj provides a blueprint for social action, and a paradigm of Hmong ethnicity and cultural opposition to the members of more dominant groups. Thus, it is appealed to by those who seek to lead uprisings directed against the state. The prophet, like Yaj Soob Lwj, like Paj Cai, is possessed by the spirit of the *Huab Tais*, Tswb Tchoj, whose imminent arrival he may announce. In proof of this supernatural sanction for subsequent action, a form of writing is revealed to the prophet in a dream or vision, which is presented as the original Hmong writing, lost owing to the oppression of the Chinese. The phenomenon of possession by the spirit of the *Huab Tais*, which Bertrais (1978, p. 19) describes as 'like a hypostasis' (the embodiment of a personality), continues to recur, as is shown not only by the messianic movement in the refugee camps, but also by the constant press reports from Vietnam that the Chinese have disseminated rumours of the birth of the Hmong 'King' in order to attract Vietnamese Hmong on to the Chinese side of the border.[14] I have also come across these rumours in Yunnan.

In the notion of 'hypostasis', Bertrais (1978) may have been thinking of the mystical re-enactment of events from the life of Christ found in the medieval mystery plays of Europe, or the notion of personal re-enactment found in Thomas à Kempis's *Imitation of Christ*, where it was felt that through imitation of the Lord's actions one could in effect 'become' Christ in a spiritual sense.[15]

One needs to distinguish this sort of 'hypostasis' from the technical aspects of ritual action, or the ritual aspects of technical action, which Worsley (1968, p. 14) rightly points out are united in the notion of 'work', since it represents an identification with a figure who stands 'outside' historical time to an important degree. Among the White Hmong, other examples of this hypostasis include the way the shaman, in trance, 'becomes' Siv Yis, the premier shaman, and may refer to himself as such, marrying couples are identified in song with the first marrying couple, and the deceased with the first deceased, during the course of wedding ceremonies and funeral rituals. It is at such moments that the course of history is interrupted; a knot has been encountered as it were, and we see how mythology may 'freeze' time by imposing its own, timeless structure on the course of events, as the unconscious does on personal experience (Levi-Strauss 1963b).

To return to the legend of Tswb Tchoj, rather than the kinds of social action it may validate, the most persistent underlying motif of the story is the *dubiety of paternal ancestry* it so clearly expresses, not only in the fact of virgin birth itself, or in the riddle in which the Hmong boy is told that he has no father (and of which so much is made in the oral telling), but also in the repeated images of violence and destruction directed against the mother. It is significant that where the Hmong has lost a father, the Chinese has lost a mother, and in Chapter 9 some aspects of the Hmong kinship system, and the suggestion that the Hmong patrilineal kinship system was adopted from the Chinese (Ruey Yih-fu 1960) are examined from this point of view.

One might be tempted to conclude that the legend provides evidence of the adoption of patrilineal kinship, and therefore an earlier matrilineal or bilineal organization, were it not that Eberhardt (1982, p. 50) has pointed out that all Chinese clans claiming sovereign authority, as most larger clans did during the Chou dynasty, had to prove their descent divine since the Emperor was conceived of as the Son of Heaven, which they invariably did by eliminating the ancestral father and retaining the ancestral mother. This was particularly true of the Han and Manchu dynasty clan legends. Could it be that the Tswb Tchoj legend, although now adopted by Hmong from all the different clans, originally represented such a clan origin myth?

The orphaned status of the hero is particularly important from this point of view, since it does to some extent allow him to transcend the bonds of affiliation to a particular clan, and represent the kind of political unity among the Hmong which is often seen as marred by clan divisions. It is for the same reason that the prophets of the *Huab Tais* are themselves commonly orphans, as Paj Cai was. The status of orphan allows a leader to appeal for support to the members of different clans. Hmong in the survey site did not know Tswb Tchoj's clan affiliation, and had clearly never thought about it, but if pressed assumed that he must have adopted his mother's or mother's husband's clan name. However, an account of the legend which I received from the refugee camp declared that, indeed, the legend had formerly belonged to the Tsab clan, and that Tswb Tchoj had been a member of the Tsab clan, who had consequently maintained a prohibition on the eating of certain kinds of pork (only from fully grown boars) ever since. Dietary prohibitions are maintained by most Hmong clans.

According to the informant, since the members of the *koom haum* (the messianic movement in the refugee camps) had begun to *haum* (worship) Tswb Tchoj, all the priests of the movement had now adopted the taboo which had formerly only been observed by the Tsab clan, and maintained a prohibition on eating all kinds of pork. This would seem to be confirmed by the subdivision of Hmong Tsab descent groups into Tsab Suav ('Chinese') and Tsab Tswb.

Since one of the major ethnic insignia in the fieldwork district was, as we saw in Chapter 4, a distinction between pork-eating Hmong and Chinese, and non-pork-eating Muslim Chinese (and Seventh Day Adventists), it may be interesting to speculate here on the origins of the Tsab clan taboo on pork. The Hmong have maintained long historical relations with the Muslim Chinese of southern China, some of whom trace their descent from Kublai Khan's soldiers in the thirteenth century (Hill 1982), and several Hmong clans are said to have descended from marriage with male Muslim Chinese. Since the first part of Tswb Tchoj's name certainly means 'pig', and the latter is often pronounced as Tshoj, and so can mean 'dynasty', one cannot help wondering if the name did not originally bear some reference to the Ming dynasty (1368–1644) whose family name 'Chu' may also be transcribed as 'pig', and who prohibited the killing of pigs early in the sixteenth century for just that reason (Eberhardt 1982).

Particularly throughout the nineteenth century in China, the cause of 'restoring the Ming and overthrowing the Qing' dynasties was espoused by many groups of rebels, pretenders, and disaffected sectors of society (F.-L. Davis 1977). Even after Sun Yat-sen's Republican victory of 1911, members of these secret societies rushed to the graves of the Ming Emperors to inform them of this change in their fortunes. The Triads and Nien rebels who flourished in the nineteenth century were organized into various 'banners', like the Muslim Chinese Ho or Haw who, together with the Hmong, ravaged northern Tongkin in the 1870s (MacAleavy 1968). One has to remember that the last great Miao rebellion coincided with the rebellion of the (Panthay) Muslim Chinese in Yunnan, who established a separatist state for themselves there from 1855 to 1873 (Yegar 1972; Israeli 1978). Chinese who sought to account for the Muslim prohibition on pork charged that they had descended from men who 'herded with pigs and infidels' and had descended from a miraculous union with a pig, to the great fury of the Muslims (J. Anderson 1878, p. 324). We know that there has been a long association between the ancestors of the Hmong and Chinese rebels. For example, it was two of the last Ming pretender's generals who took refuge among the Miao and first taught them the use of the flintlock (De Beauclair 1956a).

The weight of evidence, then, would suggest that this extraordinary legend, paradigmatic of the very bases of Hmong ethnicity as it is defined in opposition to that of the Chinese to such an extent that it provides the continued legitimation for the uprisings of the Hmong against centralized authority, was originally a clan origin myth of some kind, of the type attributed by the Chinese to Muslim rebels with whom the Hmong have had long historic contacts, and from intermarriage with whom some Hmong clans descend. However, there are strong Daoist and Buddhist elements in the legend as well, and it is very unlikely that Muslims themselves would have been responsible for the transmission of what would have been a scurrilous tale about themselves. Above all, however, it remains a puzzle why the legend should fall into the generic pattern of stories about *geomancy*, and why the Hmong should have adopted such a story to explicate their own ethnicity and opposition to the state. Not only is the Chinese-derived term *Huab Tais* (or *Fuab Tais*, or *Faj Tim*) employed by the Hmong to refer to their own 'Emperor', but his prophets even use the Thai or Lao-derived term *Chao Fa* (Heavenly Lord, formerly applied to earthly princelings) to refer to themselves. In order to investigate this problem more closely, it will be necessary to inquire into the system of geomancy practised by the Hmong.

1. The Earth King is a figure encountered in the myths and legends of many neighbouring peoples, often associated with mines and mining. Many of the other themes have local parallels, e.g., the Yunnanese story of 'How the Oxen Twisted their Horns', in which a *Tai* girl discovers her father to have been an ox (Christie 1968).

2. It was not clear from the informant what was woven, although it must have been a magical fabric of some sort.

3. Here there has been an obvious confusion of names, since *Tuam* is the Hmong version of the Chinese birth-order term for the eldest son, and *Xyos* of that for the youngest or last-born. Thus China itself is referred to as *Tuam Tshoj Teb* (the lands of the great dynasties) and the lands of Indo-China beyond the Chinese frontier as *Xyos Tshoj Teb* (the lands of the lesser dynasties).

4. I owe this identification of the *Xib Nyuj* as a rhinoceros to Professor Downer of the University of Leeds (Personal Communication, March 1984).

5. Another curious denial of the physical facts of existence.

6. Hence symbolically destroying the potency of the Chinese ancestral line.

7. It is odd that the decree of heaven should be voiced at this juncture through the medium of the Chinese. In another version, a small bird voices the message. The 'Golden City' here poetically symbolizes the earthly as well as celestial realms.

8. The doubling of the figure of Tswb Tchoj, in the elder and younger brothers, represents a submerged theme of sibling rivalry emphasized by another informant (see Chapter 6) to distinguish the Hmong and the Chinese.

9. It is significant that Tswb Tchoj should be depicted as carrying a book and a pen at his birth: Paj Cai was described to me in the same terms by another informant in the village. This provides further evidence of the mystical script which messianic leaders adduce in evidence of their sanction by the spirit of Tswb Tchoj.

10. Stephan Feuchtwang, Personal Communication, January 1984.

11. Viz., Eberhardt, Cycle No. 172 (1968, p. 226). For other examples, see Freedman (1968) on geomancers' needles, arrows shot to kill the reigning Emperor, and muddy duck ponds.

12. Probably his clan name, Xyooj.

13. Man refers to the Yao. The Nung and Tho are separate groups.

14. For example, *The Guardian*, 9 January 1981: 'China pushes ethnic cause in border row with Vietnam'. See also Vietnam News Agency reports in 1978 that the 'Hoa' Chinese were spreading rumours that the Hmong King 'is now very old and is longing for all his sons and daughters to return', *Far Eastern Economic Review*, 1 September 1978.

15. Thomas à Kempis, *The Imitation of Christ* (J. M. Dent and Son, London, 1928).

PART IV

8

The Place of Geomancy

IN the first place, it is remarkable that a geomantic system for the siting of ancestral graves usually seen as uniquely 'Chinese' should be practised by the members of a minority culture who do not, themselves, practise the custom of 'double burial' with which the system is in southern China so closely associated (Freedman 1968). In the second place, it is remarkable that the downfall of the messianic hero of that minority culture, considered in the previous chapter, should be attributed to this system. Geomancy among the Hmong was first noted by Lemoine (1972a, p. 99). No study of its application or significance has yet been made, however, and although it is difficult to schematize the principles of this unwritten system unless one is actually in the mountains where they are applied, the following case-studies of its application, together with legends and a map associated with the origins of the system, must serve as preliminary steps towards such a study. It will be found that elements of the legend which depicts an ideal-ized representation of the system of geomancy bear pointed resemblances to elements of the Tswb Tchoj legend considered in the previous chapter, and this is considered in the remainder of this chapter.

Several cases from within the survey site illustrated the importance which was attached by villagers to the geomantic system as a means of explaining fortune, misfortune, and illness. Here I shall refer to five of them.

Applications of Geomancy

Nkaj Suav, the richest man in the larger settlement, had acquired his wealth over the years largely by being sharp-witted in choosing the best locations for his crops and alternating them in accordance with the vagaries of the weather. He was excessively hard-working, sleeping little, and with every second of his time employed even at home, for example, in rethatching his roof or weaving back-baskets and winnowing panniers. He also had the physique and constitution of a bull, and so was the usual village guard, patrolling its outskirts throughout the night. The other villagers, however, offered two contrary explanations for his wealth, which were not wholly of a pragmatic nature.

The first of these was that he had been particularly fortunate in the choice of a burial site for his first wife, whose early death had enabled him and the new wife he then took to accumulate fortune and prosperity. The second was that, at the time when many Hmong had flocked up to the province of Chiangrai to join the Hmong *Huab Tais* they had been led to believe had arisen there, he had remained behind and had thus been enabled to take over the best land. While the latter explanation referred to an occasion of a supernatural nature, it was a practical one, nevertheless, and pointed towards his own shrewd and practical nature. The first explanation, however, while not denying his pragmatic outlook (since the system of geomancy is regarded as a thoroughly pragmatic one), nevertheless involved the geomantic system as an explanation of fortune which I felt to be of an essentially supernatural nature. One should note here the importance of the geomantic system in explaining the allocation of a scarce resource—*land*, for which the villagers were naturally in competition.

This was also the situation with the case already referred to (in Chapter 2) of the notable shaman and ex-member of the CPT, Tsav Pheej, a *kwvtij* (lineage) relative of Nkaj Suav and so also a member of the dominant Thoj clan in the larger settlement of Hapo. Here the Xyooj inhabitants of the old village of Nomya, which no longer existed, had called him in to account for the number of stillbirths and miscarriages which had occurred. He had attributed these misfortunes to the alignment of the mountains encircling the village, which in this case were too close together. This had allowed a malevolent *dab qus*, or forest spirit, who inhabited a nearby spring from which the stream which watered the village ran, to plague and 'intimidate' the villagers, and to 'afflict' their women. The only solution he advised was relocation of the village, and the villagers consequently moved out of the village leaving the shaman free to cultivate the fields around the village, which proved to be particularly good opium land. Some resentment was felt towards him for this decision; although I never came across a case where the yields or fertility of the land were said to be directly affected by geomancy, in this case it was felt that if the malign influences had really been as powerful as he had claimed, he should have suffered some misfortune himself while cultivating the land. Here, however, one can again see the importance attached to geomancy as a means of explaining fortune and misfortune, as well as its relevance to the allocation of and contest for land. Metaphorically speaking, alteration of the 'prospects' of the landscape (by moving the village) was identified with alteration of the 'prospects' of its inhabitants (by avoiding future stillbirths and miscarriages). The geomantic system *is* thus 'essentially turned towards the future' (Freedman 1966, p. 102).

A similar case where geomancy was employed to account for personal misfortune came from a village not directly within the survey site, which was clearly a post hoc 'elaboration' of belief. A villager well known for his violent temper had recently taken a second wife in a love-match, whom he much preferred to his first, far plainer wife. The two wives were always quarrelling, and unlike the usual practice where a second, younger wife is expected to work in the fields so that the older one can stay at home, he insisted on his first wife continuing to work while his new bride rested. While clearing fields with his first wife, a quarrel had broken out over his second wife, and he had savagely reversed his truck into a tree which had fallen on her and injured her fatally. The quarrel must have been intense as he had then driven off at top speed and overturned the truck, nearly killing his ten-year-old son who was with him, and injuring himself quite severely. When he was accused of murder, he used a geomantic idiom to express his differences with his first wife and her family, claiming that her ancestral spirits had provoked him to run amok because her brother's front door was directly opposite his own. Spiritual influences are believed to travel in straight lines, and direct oppositions are therefore usually avoided in geomancy, as Freedman's (1968) observations on the alterations which had to be made to the doors of flats in high-rise Singaporean buildings showed.

Another case I was able to investigate illustrates the essentially *conservative* function of geomancy in acting as an idiom for social change (Feuchtwang 1974).[1] And here one is reminded of the many cases in southern China, where the system is most prevalent, recorded by Eitel (1873), Edkins (1872), and others, where the railway tracks proposed by foreigners, or the height of their church steeples, were met by objections that they would disturb the geomantic topography of their various locations. Indeed, in these sorts of examples one can see very clearly the contest for resources, such as land, which was at that time taking place with regard to the members of two different *ethnic* groups, the Chinese and the many Europeans who were in positions of economic superiority and able to exploit natural resources.

In this case, local development workers had persuaded the Hmong of a particular hamlet to abandon most of their swiddening of dry rice and poppy and had aided them at some expense to establish terraced wet rice fields on the lower slopes, explaining to them the 'quasi-ecological' theory (which I have criticized in Chapter 2) that their deforestation of the higher slopes for swidden agriculture increased the run-off rate of water down the mountainside, which contributed to floods and drought in the lowlands below. However, later the villagers had abandoned most of the terraced rice fields, and reverted to clearing the higher slopes for swiddens. When called to account for their actions, the Hmong denied that the floods and droughts in the lowlands were caused by their own actions, and attributed the excessive surface downwash which had occurred to the movements of a giant dragon (*Zaj*) which inhabited the mountains.

Although this view did not make much sense to the development workers, who dismissed it as a pretext for 'idleness', when we understand that here the most fundamental metaphor of the geomantic system is being employed, we can well see how the discourse of geomancy played a crucial role in articulating villagers' attitudes towards externally induced developments, natural misfortunes such as floods and drought, and the contest between members of different ethnic groups for control over land-use systems. For the primary metaphor in the geomantic idiom for the topography of the land is precisely that of the *dragon*, which is believed to inform all the contours of the mountain ridges and watercourses, to the extent that the Hmong *term* for geomancy, *loojmem*, or the 'veins of the dragon', refers directly to the metaphor of the dragon as an image of natural, unrestrained energy. In fact, long lines of mountain peaks do look very like traditional depictions of Chinese dragons. Moreover, as the tale of Yaj Xeeb Xeeb (below) shows, certain dragons are particularly associated with the control of the waters and rains.

Finally, and perhaps the most indicative of all, a case involving the members of a mining company took place within the survey site with regard to a seven-household White Hmong hamlet. Nine years before my arrival in the area, a tin-mining company had initiated operations in the valley beneath the slopes along two of which the village was situated. Employees of the mining company established a camp at some distance above the village near a spring from which they began to construct a watercourse to their mining concession in the valley below. The Hmong objected to this on the grounds that, by doing so, the watercourse would 'cut' the *memtoj* (veins of the slope) which channelled a continuous and creative energy (*pa*, literally, 'breath') along the contours of the mountainside on which their village was situated.[2] The fortune of the village depended on this supply of *pa*.

There are two related beliefs here. The first is that change in the landscape will bring about change in the life of its inhabitants, and the second is that such changes will be maleficient. Again, as in the case of the relocated village above, the villagers realized their prospects would be altered—and expressed this realization through the idiom of geomancy.

The villagers objected to the proposed watercourse so strongly that a special meeting had to be called by the mining company. It was agreed that the village should be relocated to a site above the watercourse, and compensation was paid to each household head by the mining company. Yet, the proximity of the mining concession led to increased interactions between the Hmong and members of other ethnic groups, and change inevitably resulted. Although the villagers could use the water-pipe as a route to lowland markets in the absence of a road, Karen and Thai from the mining camp began to cultivate some of the rice fields around the old location of the village, and even cleared some opium swiddens further up the mountain. The village, in effect, lost access to a good deal of the land it had previously cultivated, and, moreover, one of the extremely rare cases of Hmong prostitution had occurred as a direct result of the establishment of the mining concession. A young Hmong girl had moved in to live with a Thai labourer in the camp, and after being deserted by him was forced to turn to prostitution for a living. There is no doubt that, in this case, fears of the possible results of increased proximity with other ethnic groups associated with the introduction of the watercourse, which were articulated and expressed with reference to traditional symbols drawn from beliefs in geomancy, were felt to have been vindicated by the actual effects of the introduction of such changes into the natural landscape. The village had become steadily poorer over the years, and only seven households remained there. The feeling was that things would have been even worse had the village not relocated.

Geomancy is thus, for the Hmong as for the Chinese, 'motivated by anxiety, brought about by the apprehension of change, by interference or a sense of challenge' (Feuchtwang 1974, p. 243). Therefore, it does indeed delineate 'an area of *uncertainty* beyond the explanatory powers of modern science' (Feuchtwang 1974, p. 202, author's emphasis), which it has been the purpose of Chapters 2–4 to describe.

To summarize the last case, the proposed alteration of the natural landscape in the form of the construction of a water-pipe augured obvious change for the villagers through increased interaction with the members of other ethnic groups. The unease and 'uncertainty' experienced about this was solved through relocating the village. Both the unease, and the manner of its resolution, were phrased through the medium of traditional geomantic symbols and ideas. The unease was, to some extent, justified by the practical effects of constructing the water-pipe. Although it brought the opportunities of increased commercial and medical contacts with a market economy, in practice such contacts proved not to be beneficial to the villagers, since their economy could not support such contacts. The increased ethnic interaction resulted in some breakdown of kinship-based values associated with the demoralization of one woman, besides loss of control over vital resources in the form of land and water. This pattern of retreat, withdrawal, and loss of vital resources has been a constant pattern for the Hmong for some 2,000 years, which we examine in Chapter 9, and it is precisely this type of rivalry and competition which it is the supreme function of the geomantic system itself (examined by Durkheim and Mauss (1963), as a classical example of 'primitive classification', and described by Freedman (1968) as 'the most systematic statement of Chinese ideas about the constitution and working of the cosmos'), to articulate, mediate, and arbitrate.

Principles of Geomancy

Near the village, I had the opportunity to visit the ancestral burial sites of many of the villagers, whose respective fortunes were explained to me in terms of their relative alignments. Vaj Xeeb, the headman, boasted to me that he had had four sons and only one daughter, because of the favourable alignment of his father's grave, unlike that of Suav Yeeb's father, which suffered from a small, marring peak, between two mountains (Figure 6). Although some of the beliefs about particular alignments appear to vary between descent groups, the map which Vaj Suav (Vaj) of the focal village traced (originally in the ashes of an outdoor fire) provides the best illustration I know of the essential principles of the system. Sighting 'left' and 'right' from the point of view of the main range of mountains, the two primary mountains, which illustrate the male and female principles of the system, then represent secondary spurs running down from the central massif, with the direction of sunrise to the left, and the direction of sunset to the right. It is most important that the male mountain should be higher than, and encircle, the female one, since if this is so it will benefit the male, as opposed to the female, descendants of the deceased. It was explicitly stated that it was more desirable to benefit male descendants than female

FIGURE 6
The Ideal Burial Site

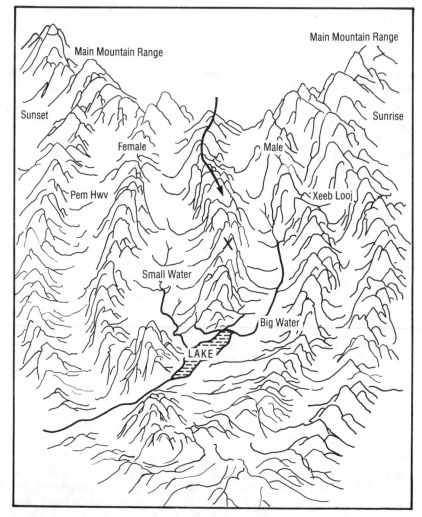

descendants because 'the sons will stay in the household and revere the ancestral spirits, while the daughters will leave the family and move away'.

The lake shown in the diagram is the site of royal rebirth, and is referred to as the *lub pas zaj*, or 'dragon's pool'. Many other Hmong stories are told of such lakes, which are inhabited by dragons. The old Dragon who rules the waters which surround the earth and cover it beneath the heavens, and is associated with fertility motifs throughout South-East Asia, is a figure met with in many Chinese stories. It was probably the disappearance of the rivers every dry season which first led to the idea of an underground cavern in which, together with the old Dragon rulers of earth, the spirits of the dead dwelt for the winter, known as the 'Yellow Springs', which became a kind of hell or purgatory in later theological formulations (Loewe 1982,

p. 34). Although most differences of opinion concern the precise location of the burial site along the central spur, there should be two watercourses to each side of it, leading into the lake.

The course of the *loojmem*, or dragon veins, which convey the breath of the dragon, is down this central spur, and the corpse must be buried at a certain time of day, usually in the later afternoon, if the *tus plig* (soul) of the deceased is to travel with it. Preferably this should take place, according to the twelve-animal calendar which orders hours as well as days and years (Coedès 1935), in the time of the tiger, since this is a 'strong' time, and never in the time of the chicken, since this is a 'weak' time. The *loojmem* usually 'flies out' at night (as may sometimes be observed in electric flashes over the mountains) and 'returns' during the day. However, since 'as man's evening is the morning of the spirits, so man's morning is the evening of the spirits', the soul of the deceased must go with the *loojmem* in the *spiritual* morning. Thus the *plig* of the deceased will 'catch' the *loojmem* before it 'flies out' at night, and 'ride' it, as one 'rides a tiger', to the place of rebirth.

In terms of the very clear analogy between physical and natural imagery expressed in the system and illustrated in the diagram, it was interesting how the basic principles of the system, which did not differ in essentials from those of the geomancy practised by the Chinese, would invariably be illustrated with reference to parts of the body. Two arms would be held out to represent the 'dragon' and 'tiger' mountains of the left side and the right side, with the burial located at the centre of the chest, or the three middle fingers of a hand would be extended to illustrate the relative heights of the three respective spurs. The system is a deeply internalized one, intersecting with the individual life-cycle at critical junctures, such as on the establishment of a house or the death of a relative. Besides determining the location of village and burial sites, and acting as an explanation for relative fortune and misfortune, the system was physiognymic in aspect and pragmatic in purpose. Thus, the peak of Nomya was pointed out to me as defending the village against thieves and officials (mentioned in the same breath). Thieves and officials might be able to come into the village, but on their approach they would fall out among themselves, and forget the evil purposes for which they had come. Suav Yeeb's house was described as the 'head' of the village, laid out in the familiar 'horse-shoe' pattern of many Hmong villages, with the two sides of the 'horse-shoe' corresponding to its arms. And the physical, metaphoric aspects of the system extended beyond the village, to include Chiangmai, which was described as secure and prosperous owing to the protective embrace of the mountains which surround it to the north. Sometimes places where thunder and lightning had struck, and meteorities, were singled out as marking particularly inauspicious spots for *loojmem*, and it was also said that the head of the corpse should be turned away from the direction of the sunrise, since if the deceased should awaken and open his eyes, he would be unable to see and watch over his descendants: 'Blinded by sun he cannot see, so the sons and daughters will be poor.'

The distinction which we encounter in the legend associated with geomancy which follows this section, between the left hand representing the stronger, male side, and the right hand representing the weaker, female side, is relevant to the 'left hand–right hand' controversy considered by Hertz (1973). The same distinction is also made by the Chinese (Granet 1973). Further associations were made by Hmong in the village, moreover, between the left hand as the 'resting' (and, therefore, stronger) side, and the right hand as the 'moving' (and, therefore, weaker) side, which were illustrated by the fact that while it is the *left* hand which holds an object one is cutting, it is the *right* hand which cuts it. Hertz (1973) does perceptively remark that it is the right hand which 'takes', the left hand which 'holds', and this does indeed seem to be a universal, although as the Hmong and Chinese cases show, the secondary associations with strength and sexuality do not remain consistent.

However, that the Hmong opposition between the staying characteristics of the stronger left hand and the moving characteristics of the weaker right hand is importantly connected with the system of patrilineal kinship, is shown in the version of the legend of geomancy given below by Xeeb Thoj (Hapo, May 1982). Here, the sibling rivalry which Feuchtwang (1974) has shown to be a primary element of the geomantic system is depicted in terms of the rivalry between *a brother and a sister*, and the reversals suffered by the brother's line, even after he has beaten his sister in a race to the dragon's pool, are attributed to her curse, and her *looking behind* her. In order to redeem this state of affairs, it is necessary for the daughter not to look behind her as she leaves the home and crosses the mountains to become a daughter-in-law:

The Brother and Sister who Contested Loojmem

What I speak of here is the foundation of learning about the *memtoj*, the veins of the slopes, before going on to sight the *loojmem*, the veins of the dragon. The beginnings were like this. When Heaven and Earth were created by the *Huab Tais*, the Emperor, of the heavens, there were mountains and valleys, peaks and cliffs, bamboo woods and grassy wastes. But these mountains and valleys which were created upon earth were not all of the same kind. Below I will explain about *xeeb looj* (the Azure Dragon) and *pem hwv* (the White Tiger), the couple who were the origins of it all, long, long ago. The elders have passed on the history of it, which I give here for all to know.

Once there was a brother and a sister. The sister was the oldest, the brother was the youngest. They were both making believe that, when they were grown up, they would be able to study and become very wise. But the sister asked her brother, 'Once you have become very wise, what are you going to do with it?' And he replied that he was going to be the Emperor who guarded the Ancient Dragon. So the sister said, 'No, I will be the Emperor who guards the Ancient Dragon. Since you are the youngest, you can be a King guarding the people. I am the eldest, so you must let me be the Emperor guarding the Ancient Dragon.' But the brother answered, 'You're only a woman. Your body is for carrying water and bearing children. You cannot govern heaven and earth.' And his sister, boiling with fury, answered him back, saying, 'What are you talking about? You show me how you're going to be the *Huab Tais* guarding the Ancient Dragon—I'd like to see you try to learn anything at all!' So the brother said, 'I can learn a magic word of truth; if you can learn any wisdom at all, show me.' And the sister replied, 'I can learn a riddle from the language of creation.' And so they quarrelled, about which one of them could go to be the *Huab Tais* and govern the Ancient Dragon.

Eventually the sister said, 'Oh! Let's go together and test to see which of us can learn wisdom the best, and who will become the *Huab Tais* in charge of the Ancient Dragon.' And the brother said, 'All right, but since you're the oldest, I'll let you try first.' The sister inscribed four characters and threw them into the four branches of heaven (in all directions). Then the heavens darkened, and rain fell lightly. But the boy inscribed a single letter and threw it into the centre of the heavens. A gust of wind and a tempest blew up, with thunder, rain, and lightning.

So the sister said, 'Oh well, if it's like that, never mind. Let's come back tomorrow and try again.' And on the next day they went to try again. The sister said to her brother, 'Each of us will have to carry ten measures of copper and ten measures of iron, and whichever of us is able to carry them, will go to be the *Huab Tais*.' And the brother said, 'All right. You're the oldest, so you go first.' The sister was able to carry ten measures of copper on the left side and ten measures of iron on the right side. But so could her brother. So she said, 'Tomorrow we'll have to compete behind the Old Dragon's house. And if I still can't beat you, you'll be the winner.' So on the fourteenth 'cow' day of the waxing moon, they went down to the shallow valley, and that was where they contested. When the couple were below the dragon's pool, where the two great mountains meet, the brother asked his sister, 'How are we going to compete for wisdom here? You decide.' And the sister said, 'This is the last time we shall try, and if I cannot win, you will be the *Huab Tais* who manages the Ancient Dragon. And if you cannot win, then I will be the *Huab Tais*. I am the eldest, so I will say what should be done. You shall go to one mountain, and I shall go to the other one. You are the son, so you go on the right-hand mountain, and if you are the first to get to the top and can find the *memtoj* first, then you will take the mountain on the left-hand side to guard you. If I am the first to find the *memtoj*, I will take the mountain on the right-hand side to guard me. We two will contest together to see who can be the first to get down below this dragon's pool, and whoever is shall be the *Huab Tais* who guards the Old Dragon.' And the brother said, 'All right.' So the two, brother and sister, struggled together to mount the slope and contest the *loojmem*. As the brother was a man, he had more strength, and was able to get up first. As the sister was a woman, she had no strength, and could go only slowly. She had only just got to the mountain on the right-hand side when what? Her brother had beaten her, and was riding along the *memtoj*. The heavens darkened, night fell, a tempest blew up with thunder and lightning, and the peak of the mountain slid down in a landslide as he rode along the *memtoj*, turned around the head, then returned downwards along the way. The sister thought she would come down the right-hand mountain to block the way, but she had just turned round to take a small jump when she caught sight of her brother riding past on the *loojmem* with the slopes collapsing around him, churning up the land like water. So the sister called on Pem Hwv to help her, lamenting that she had lost everything, and cursing the side of the left-hand mountain. And the brother called on Xeeb Looj to help him guard the many dragons and the Dragon King, for nine stages and eight generations, so that he would be the *Huab Tais*.[3] And he rode the *loojmem* until he got down to the dragon's pool where the old dragon, little by little, raised its head above the waters, and then suddenly he jumped on it and rode it astride its neck behind its head. So that thus he was able to become the *Huab Tais* who was in charge of the Ancient Dragon for ninety-nine years.

Then the Old Dragon said to Lij Xeeb, 'Lij Xeeb, you are only nineteen years old and yet you have governed me for ninety-nine years already. Return and rest for three years, and then you can come back to govern me again. I beg to say that although you are the *Huab Tais* who has governed me, the Ancient Dragon, for ninety-nine years, you have governed me so severely that this old Dragon hasn't had a single full night's sleep. So you return now, and if you can stay there for three years, then you can come back and be the *Huab Tais* again and govern me forever. But if your luck runs out and you cannot stay there for three years, then you will die, and only be the *Huab Tais* as much as you have been until now.'

And Lij Xeeb thought and thought, and eventually he said, 'All right, but if I am lucky and I can go back to rest for a full three years, I shall return to control you until the very end, while if it should be that I have no luck and should die, I will *still* return to be the *Huab Tais* and govern you in the spiritual world for nine generations.' And the Old Dragon agreed. So Lij Xeeb went back home and lived there for another two years. But before the three years were up, his luck ran out. His body became weak and listless, and he told his sons and grandsons that if the time had come when his luck should be ended and his licence of life run out, and he should die, then they should be sure to seek out a good place to put him in (to bury him in), so that he could return to be the *Huab Tais* who guarded the Ancient Dragon. But he did not tell them of the place where he should be put. And then he died.

His children summoned all their relatives and elders to a grand meeting to discuss what should be done. Everybody agreed that since he had once been the *Huab Tais*, he should be taken to be buried near a very high mountain ridge, and they should mark his grave with stones. And so they took him up on to a high mountain ridge, and built up stones together upon the place. But whatever his descendants did, after three years he still had not arisen. At that time there was one Hmong, Yaj Xeeb Xeeb, perambulating the heavens and earth, and so they decided to go and ask him why it was that, whatever they did, after three years, their father had still not arisen.

And Yaj Xeeb Xeeb said, 'It's not because of what you have done, it's all because of your Txhiaj Meej [the spirit of wealth and richness who guards the door of the house], and because your mother and father have not good *loojmem*. It is because they have not good *loojmem* that they are unable to watch over you and protect you. Your Txhiaj Meej being weak, does not arise to guard your door, so the spirit of breaking things has penetrated and forced a way through, so that all is lost and broken, and whatever you do, your father will not arise.'

When the children had fully understood the basis of what Yaj Xeeb Xeeb was telling them, he continued, 'Oh! I know what you should do. I, Yaj Xeeb Xeeb, have lived for nine stages of heaven and eight stages of earth. For a full three years now the Old Dragon has not managed the waters and rains well for the people of this world. The crops have not been good, and the people have fiercely reviled the Lords of Heaven, who had despatched me to encircle the heavens and the earth to find out why this has been so . . . but so far I have not been able to. Still I can help you. Catch a white chicken and make incense and three pieces of money paper for me to banish your spirit of breaking things. Then catch a red cock for me to raise up your Txhiaj Meej, and in three days I shall arrive. Open your door so I can help you to sight the *loojmem* for your mother and your father, *loojmem* that will be good.' And so the sons searched out chickens and incense and paper for Yaj Xeeb Xeeb to raise up their Txhiaj Meej.

Then Yaj Xeeb Xeeb told them, 'This breaking spirit, is all because of your father's sister's curse. This spirit is haunting you, and will ruin all that you attempt, so that you will have no wealth or anything else, whatever you do. Now I have caused your Txhiaj Meej to rise up, which is to say that all is good. Now, if you have one young girl to go across the foot of your mountain to become a daughter-in-law, and you tell her *not to turn around and stare behind her* (like the aunt), then this spirit of breaking things will be unable to trouble you any longer.' And that was how Yaj Xeeb Xeeb taught them.

Three days later he came with them to sight the *loojmem* for their mother and father. Yaj Xeeb Xeeb was able to find the right place for their father, and he told them that that was where they should bury their father: 'Take his bones to put them in this grave in this place, and that will be good. When you are digging the grave for your father's bones, take branches and stick them in so that water will penetrate down to the bones and all will be released, and when you visit the grave after three years, you will see something good. If you bury your father in this grave here, then your mother and father will return to be the *Huab Tais* and guard the Ancient Dragon for nine tiers, and you will be Kings governing the populace for nine generations.'

And they asked Yaj Xeeb Xeeb how long that would be, and he said in three years' time. And at that time they really did become kings and rule over the land. And now when people die, they go to sight the *loojmem* for them, of which this is the basis.

In terms of the diagram (Figure 6), the brother and sister start out at opposite sides of the large mountains' meeting place below the dragon's pool, ascending up the right- and left-hand sides respectively (sighting from the top of the diagram) to the mountain ridge at the top, where they turn around the 'head', the peak of the mountain, to descend southwards towards the pool. With the landslide, the right-hand mountain slid towards the centre, so that the brother was able to travel directly down towards the pool. The two brothers in the next version follow a similar route, struggling for supremacy below the dragon's pool, to see who would be able to go first.

The story is interesting ethnographically because it clearly relates the system of geomancy (which also orders the siting of villages) to the importance of rainfall and irrigation, and thus explains the introduction of the Dragon King who controls the waters. It also provides a precedent for the custom of *tsa txhiaj meej*, or raising the Txhiaj Meej, who is symbolized by silver coins pinned to a red cloth above the lintel of the front door, which is performed every New Year by the household. The sibling rivalry with which the geomantic system is so often associated is clearly expressed, and the left male–right female analogies remain constant.

One might have expected to find the normal Chinese oppositions reversed for the Hmong, as they are in so many other ways. For example, the sun is depicted as female and the moon as male, in Hmong descriptions, which is the opposite of the Chinese, while the normal Chinese word-order which places male before female names is invariably reversed in Hmong. However, the Hmong association of the left with the male, the right with the female, as also the *yin* with the spiritual and heavenly, the *yang* with the material and earthly, remains consonant with Chinese dualities. We see this in general terms as resulting from the way in which the opposition between Hmong and Chinese ideological categories has itself been internalized within the Hmong classificatory system, as we show has been the case with ethnic and kinship categories in Chapter 9.

The legend which follows this section shows that it is believed to have been Khoo Meej (described as a star who came down to earth), who first taught the Hmong the uses of geomancy, according to which there were strong and weak places in the mountains, just as there are strong and weak times of the days and years. Khoo Meej must refer to Zhuge Liang, or Kong Ming, the famous Three Kingdoms strategist and soldier under Shu Han, a symbol of resourcefulness and wisdom in Chinese folklore who, however, died early, after fleeing to Sichuan where a kingdom famous for its benevolence towards the common people was established.

The soul of an ancestor fortunate enough to be buried in such an ideal site as is described in both these accounts would follow the *loojmem* down the central spur of the mountain until it reached the *lub pas zaj*, or Dragon's Pool, where it would become a 'King in Heaven'. And, in a phrase which expresses the very essence of geomancy, it is added in the following account that 'When the father is King in Heaven, the sons will be Kings on Earth'. Thus it was to this system that Tswb Tchoj, the hero of Hmong culture, made appeal (Chapter 7).

However, just as the efficacy of shamanism was described as having deteriorated since the time of Siv Yis, the premier shaman, so too the knowledge of geomancy was said to have become gradually more and more imperfect over the course of time. Not only was it now impossible to find such suitably shaped and sized mountains as had existed when the Hmong lived in China, and might still exist there, as would ultimately assure one's line of royal birth, but it was also believed that such knowledge inevitably diminished with the passage of time, as it was transmitted from generation to generation, since everyone kept back a little of what they knew, even from their sons, for themselves, until at last there would be nothing left.[4] This 'theory of deteriorating knowledge' is interesting as a secondary elaboration because it employs a *pragmatic* explanation for the failures of *ritual* practice which make the beliefs associated with those ritual practices logically unfalsifiable. It also seems to be shared by other peoples of the region (Gilhodes 1909), and precisely the same theory is reflected in the Hmong stories accounting for the loss of literacy considered in Chapter 6.

Thus, the past is often referred to as the *lub keeb*, 'foundation' or 'basis', or *hauvpaus*, 'roots', 'origin', or 'source' of things, by contrast with the present, which is referred to as a time of 'flowering', or *lub ntsis*, 'tips' or 'shoots'. Here the image of 'flowering' does not carry the same connotations of fulfilment and abundance it might in a European context, since strength is thought of as residing in the roots of things, and 'flowering' has something of the resonances of 'going to seed'. In the light of the notions about sexuality examined above, it need not surprise us that sons may therefore be called the 'roots' of the family, while daughters remain its 'flowers'. In an important way, this relates to the ways in which the phenomenon of 'hypostasis' considered in Chapter 7 expresses a particular denial of the passage of historical time. The 'deep structure' of historical time is provided by those things which are permanent, such as *kevcai*, or 'customs' (literally, 'the ways of prohibition'). Change takes place only at a superficial level, by a constant process of 'sloughing off' those things which are inessential, as daughters are to the continuation of the family line. This is partly why I have described the transitoriness of the village as arising not out of the dialectic between one stasis and another, but between stasis and change itself. It is also why it is important that the most recent diaspora of the Hmong to the West should be phrased by the oral tradition in terms of the previous exodus out of China, as we saw in the conclusion to the tale of Tswb Tchoj (Chapter 7), and will find again in the consideration of the refugee situation in Chapter 10.

Origins of Geomancy

When I asked about the origins of geomancy, however, I was surprised to receive a conceptual map of the system in the form of an account of the original fission between a pair of brothers described as 'Hmong' and 'Chinese'. Here it should be emphasized that it is rivalry between male siblings which is usually associated with the geomantic system. The account given earlier of the rivalry between a male and a female sibling is highly unusual, and may reflect the greater prominence of women in Hmong society. The following account by Vaj Suav Vaj (Nomya, October 1981) is of crucial importance because it not only illustrates the male sibling rivalry often associated with geomancy, but attributes the very origins of the geomantic system (since this is what I asked about) itself to ethnic fission.

The Brothers who Struggled for Loojmem

Since, elder brother, you want to learn about geomancy, then let us begin like this. Many millions of years ago, we Hmong used to live in China. At that time, both Hmong and Chinese were very poor. Everybody liked to do good, and everybody did good things. Everyone wanted to be good, and to be strong, so they would look for the best possible places to bury their parents. Then one man found the perfect place, the best place of all, for his mother and father, and went to perform the mortuary rites for them three years after their death. The spirits of the earth and heavens, who live in the sky, told one of the stars to go down on earth and become a man. I do not know his name, but the Chinese call him Khoo Meej. And he taught everyone how to do everything such as the arts of war and the laws of the state. Then the Hmong really knew when and how to bury their parents, and could find places with good *loojmem* easily. Khoo Meej taught that both in Heaven and on Earth there were places which had strength (*zog*) and places which had no strength. There are both strong and weak places in these mountain valleys, just as the right hand toils (*khwv*) while the left hand rests (*so*). No matter which mountain valley you choose, all have both strong and weak places. When we live between two mountains, we call the one on the right the side of *Pem Hwv* (the White Tiger), and the one on the left the side of *Xeeb Looj* (the Azure Dragon). To set up a village we must know where the *loojmem* is and somebody must know how to look for it.

Once two brothers went to look for good *loojmem* for their own graves. They both found the same good place, and each of them told their sons, 'When I am a hundred and twenty years old, you must bury me here.' No matter whether they are Hmong or Chinese, they cannot be buried in the same place. But they were buried very close together. Because the older brother was richer than the younger, gold was placed under his neck, and stone under the neck of the younger brother, since after three years have passed, he whose neck has first broken will first go to be a king in heaven. And when the father is king in heaven, the son will be king on earth.[5]

The elders teach us to plant flowers on graves, and the flowers on the younger brother's grave blossomed before the flowers on the grave of the older brother. When the older brother's sons saw this, and saw that the younger brother's neck was breaking, they took away the stone and replaced it with the gold. When three years later the younger brother's sons came to visit their father's grave, they found the grave of their uncle already open, for he had gone to be a king in heaven. But they realized what bad men their cousins were—and since that day Hmong and Chinese have never got on together. The Hmong moved away from the Chinese and did not wish to speak the same language any more. The Chinese wanted the Hmong to speak Chinese, but the Hmong did not want to and departed.

We must look for certain places in the mountains: rough, jagged mountains are no good, we can use them but later the sons and grandsons of the dead will become thieves. The mountain to the left is the stronger one, that to the right the toiling one.

We have to look for good *loojmem* for our mother and father because if we don't our sons and grandsons will not be good people. When they die a good place must be chosen for them. Hmong may not be great soldiers or very rich, but so long as a good burial site is chosen for the ancestors, they will be able to live for themselves and not work for others.

There are two ways of using *loojmem*—when someone dies, or setting up a village. Houses, too, should not face each other [be in straight lines].

Figure 6 was drawn in illustration of the ideal location for which the two brothers struggled, in the above account. As in the tale of Tswb Tchoj, again we find that it is the belief system associated with the practice of geomancy which in the most fundamental way seeks to account for the rivalry and competition between two different *ethnic* groups. Feuchtwang (1974, p. 216) has isolated the rivalry between *siblings* as an essential component of the geomantic system, which distinguishes different lines of descent according to the location of the burial site of a common ancestor. By allowing a focus for individuation and social conflict, the system thus provides an important counter-weight to the system of ancestral worship. With the Hmong we find an extension of this rivalry between siblings to cover the rivalry between clans, and different ethnic groups.

Thus, the entire political conflict in Laos between Hmong supporters of the Pathet Lao and Hmong supporters of the Royal Lao government may be phrased in terms of the geomantic system. For while Tub Npis, a member of the Lis clan, became Minister of Justice in the government of 1960, Faiv Ntaj, his mother's brother and a member of the Lauj clan, became Vice-Chairman of the opposition Pathet Lao. Political support among the Hmong in Laos closely followed these cleavages of clan. The feud between the Lis and Lauj clans, said to have originated from the mistreatment of Tub Npis's mother by his father, and her subsequent suicide, was attributed to the fact that her father and her husband were then buried on *opposite sides of the same mountain* (the most fundamental opposition available, the *yin* and *yang* of the rainy and dry sides of the mountain, respectively). This was known, and talked about, in the village. The extension of the sibling rivalry which characterizes the geomantic system to clan and ethnic differentiations, together with the greater importance which seems to be accorded to the burials of wives and mothers by the Hmong, may represent one of the fundamental differences in the system as it is practised by the members of a communally ordered society, such as that of the Hmong, rather than by the members of a highly stratified society, such as that of the Chinese. Indeed, the uses of geomancy to articulate inter-ethnic rivalry may represent a more fundamental usage of the system than its use to distinguish different descent lines within the same ethnic group.

In the above account, we are inevitably reminded of the informant who claimed, apropos of the story of the loss of books, that once the Hmong had been as *older brothers*, but that now they were in the position of *younger* brothers. The use of sibling terms to distinguish the Han Chinese from the members of national minorities is common in present-day China, as Eberhardt (1982, p. 126) points out. Thus, the Chinese refer to the Miao as their younger brothers, meaning that they should honour and obey their 'older' brothers. We find striking confirmation of this in the following story from southern China (Graham 1954, p. 27):

In the earliest times the Chinese and the Miao were one family. The Miao was the older, the more powerful, and the more respected brother, and the Chinese was the younger. The parents died and were buried. The brothers separated and lost all traces of each other. They both commemorated their ancestors at the same grave, but at different times, so they did not meet.

The younger brother, the Chinese, worshipped later in the year, but finally the older brother noticed that somebody was worshipping at the grave of his parents. 'Who is doing this, and for what reason?' he asked. Then he began to watch, and caught the younger brother. A quarrel ensued. They did not recognise each other, so each blamed the other for worshipping at his ancestral grave. Instead of fighting they went to law about it. The official asked the Miao, 'What evidence have you that this is your ancestral grave?' He replied, 'I have buried a millstone, a certain distance to the right of the grave.' He asked the younger brother the same question. He answered, 'I have buried a brass gong a certain distance to the left of the grave.'

The official sent men to dig and they found both the millstone and the gong. Then it became known that the two were brothers. But in the centuries that followed the descendants of the two brothers grew apart and forgot their common ancestry, and so the Chinese have forgotten it altogether. Moreover, the Chinese descendants have grown more and more powerful, so that the Miao are now the younger and weaker brothers and the Chinese are the older and stronger brothers.

What is remarkable about the geomantic tale of the two brothers I was given in the village, however, is that the two brothers should not be seen as Hmong or Chinese: the legend refers back to a time of undifferentiated ancestry, prior to the fission between 'Hmong' and 'Chinese' communities, both of which are seen as descending from a common ancestor. Further questioning confirmed this. In fact, I was often told by Hmong that Hmong and Chinese descended from the *two wives of a single common male ancestor*. In a polygynous kinship system, as Fortes (1949) has shown, different lines of descent may be conveniently distinguished by reference to *maternal*, rather than paternal, ancestors. Thus, it was often said that the Hmong and Chinese had *sib ncaim* (separated, divorced) and, moreover, as we shall see in Chapter 9, a Hmong may ask another Hmong on meeting whether he is 'from the Hmong or Chinese mother'.

In a similar way, with the geomantic system itself, through which such ethnic differentiation is articulated, no consciousness was ever shown that the system was not at least as much a Hmong, as it was a Chinese, one. Although informants would often testify to the superior *knowledge* the Chinese had of the system, this never implied that the *system* or the elements from which it was composed was specifically Chinese, and was rather a tribute, one feels, to the ideas so clearly expressed by the Hmong legends about the system, that the system had throughout history been constantly manipulated by the Chinese to assure their own advantage over the Hmong.

For there can be no doubt that such legends allude to the long historical process in which the Hmong have been marginalized by members of more dominant ethnic groups. As we have seen in the case-studies drawn from the focal village, the geomantic system itself is importantly about the contest for scarce resources, particularly of *land*. Even the burial sites contested in legendary geomantic formulations represent areas of *land*, possession of which would assure a lineage of sovereignty over *land*. And it was for land that the long historical process of struggle and competition in southern China took place between the northern Chinese settlers and members of indigenous minority groups such as the Miao (Wiens 1954; Fitzgerald 1972; Moseley 1973).

If the system is indeed a common one, elements of which were shared by the members of different ethnic groups, then it is not surprising that it should be in this system that the Hmong should seek to express the repeated failures of their attempts at sovereignty, represented by their messianic hero, since the system functions, above all, as a medium for the articulation and expression of contests and rivalries which may be of an inter-ethnic nature. We have seen how this works in the case-studies of its application in the village. And if there was indeed a time before the differentiation of 'Hmong' from the 'Chinese' ethnic categories, then it is not remarkable that geomantic tales about social differentiation should refer back to such a time. If, however, one continues to see the geomantic system as uniquely or specifically a 'Chinese' one, then the mystery remains: why should the Hmong seek to explain their ethnic differences from or opposition to the Chinese, in terms of it? The fact remains that, inquiring about the Hmong messianic hero who fought against the Chinese, one receives a story about geomancy, and inquiring about geomancy, one receives a story about the differences between the Hmong and Chinese.

Analysis of Structural Elements

Apart from their generic resemblances, there is one theme which remains constant in both the legend of Tswb Tchoj and the male sibling story above associated with geomancy, and this is the *manner* in which the Chinese assure the subjugation of the Hmong. Since it is most often through precise differences of funeral and burial ritual that different Hmong descent lines are distinguished, it will be as well to examine this further here, in conclusion to the chapter. For while in the geomantic legend the gold of the (ancestral Chinese) elder brother's grave is exchanged for the stone of the (ancestral Hmong) younger brother's grave, in order to assure the subjugation of the latter, in the messianic legend of Tswb Tchoj iron is not sufficient to assure the subjugation of the Hmong, and has to be replaced by stone. Here the oppositions between Hmong and Chinese, and stone and metal, remain constant, although their combination varies.

In a second version of the Tswb Tchoj legend, collected from a Xyooj villager, *stone* in the grave of the Hmong was insufficient to ensure that he would not be reborn, and had to be replaced by *iron*. Thus, the opposition remains constant, and it does not matter if the informant in the first version of the Tswb Tchoj legend had become confused (as he was with regard to the birth-order terms for Tswb Tchoj), since the opposition between metal and stone expresses itself so clearly. Moreover, in all three cases, iron is seen as inauspicious, since it is not suggested, in Xauv's version of the Tswb Tchoj tale in Chapter 7, that iron was *auspicious*; indeed, the Chinese buried iron in the Hmong grave for the same purpose as they then buried stone—merely that iron was *insufficiently* inauspicious.

However, in the third, abbreviated, version of the Tswb Tchoj legend attributed to Paj Cai, the messianic leader of the 1918 revolt in Laos, Tswb Tchoj is given both a slab of stone for a pillow *and* a bar of iron for his neck, while the *Mab-Suav* (Yi and Chinese: non-Hmong), who have deceived Tswb Tchoj into agreeing to return with them to 'sleep' for another 180 years, give themselves a *wooden* pillow for their grave (Yaj Txoov Tsawb 1972). This too, however, ensures that they are reborn before Tswb Tchoj, who thus loses the chance of governing the land. While the opposition between Hmong and Chinese remains constant here, the third element of wood has been introduced, which is opposed to the (inauspicious) iron *and* stone buried in the Hmong hero's grave. While here the opposition seems to be between perishable *wood*, and other more imperishable materials, it is still the *metal* which remains depicted as the inauspicious material which prevents the rebirth of the Hmong hero who is to become Emperor.

This is because it is believed that everything that is buried in the grave must perish if reincarnation is to occur. Corpses should not even be dressed in synthetic materials nowadays, on account of the strength of this belief, which as we will see in Chapter 10, is still prevalent among the refugees in the United States. It is because the wood perished before the iron and stone in Paj Cai's version of the Tswb Tchoj legend, that the others were enabled to be reincarnated first, and not until a dog had burrowed under Tswb Tchoj's pillow in search of a nest of rats was Tswb Tchoj enabled to arise, vowing vengeance on the members of other races who had robbed him of his birthright.

The real opposition is one between perishable and imperishable substances, in which metal remains consistently inauspicious for the Hmong, and it is remarkable that this opposition should be so invariably associated with the opposition between different ethnic groups (although technically two oppositions are involved here; one between metals and other substances, the other between perishable and imperishable substances). For we again find it in the legend which accounts for the differentiation between Hmong and Chinese siblings given earlier (Graham 1954). Although employed in a different way, here it is the Miao brother who proved his descent from the common ancestor by having buried a *millstone* close to the grave, while it is the Chinese brother who likewise proves his descent from the same ancestor by having buried a *brass gong* close to the grave.

It is most striking that in the Hmong legends of the Deluge, which have been examined most fully in Lemoine (1972b), precisely the same oppositions should emerge. I will not rehearse here the many variants of the Deluge which have been collected, but note that their usual form describes the way in which a brother and a sister (comparable to Fu Hsi and Nu Kua of Chinese mythology) survive the Deluge and marry after a series of ordeals which classically involve each of them rolling the two pieces of a stone mill (referred to as the male and female pieces in Hmong) down opposite sides of the same mountain. Their coming to rest together at the foot of the mountain is a sign that they must marry, but from their union results a shapeless lump of flesh, which Saub, the wise old seer who guides humanity, advises them to cut, from the various pieces of which descend the clans, humans, or, in some versions, different ethnic groups.[6]

In the classic form, however, it is the youngest brother who survives the flood, together with his sister, while his eldest brother perishes, and the manner in which this is accomplished clearly manifests the oppositions we have noted above. Thus, in an account from northern Vietnam, the elder brother perishes in a boat of *iron*, which sinks, while his younger brother survives in a boat of *wood*, which floats (Savina 1930, p. 152). In an account from southern China, the elder brother perishes in an *iron* barrel, while the younger survives in one of *wood* (Graham 1954, p. 179). Lemoine (1972b) provides a version from Laos in which the elder brother perishes in a drum of *iron*, while the younger survives in a drum of *wood*. And, finally, in an account from Northern Thailand, the elder brother perishes in an *iron gong*, while the younger survives in a *wooden gong* (Mottin, n.d.(a), p. 27). The latter provides another variant, in which the elder brother perishes in a *copper* gong, while the younger brother survives with his sister in one of *iron*. This is a notable exception to the otherwise constant opposition between the inauspiciousness of metal, associated with an elder brother who perishes, and the auspiciousness of wood, a perishable material, associated with a younger brother, who survives, and who is Hmong.

Were it not for the male–female sibling version of the geomantic tale which I have given above, one might be puzzled as to why these oppositions should characterize flood legends as well as those of a geomantic type. However, in this version we find the usual rivalry between male siblings over the site of geomantic rebirth transmuted into *a rivalry between siblings of the opposite sex, a brother and a sister*, whose three contests closely resemble the three ordeals to which the incestuous couple must usually submit themselves before they are allowed to marry.

Indeed, both the messianic and geomantic tales we have considered, and the legends of the flood, resemble each other importantly in seeking to account for the origins of different kinds of social differentiation. Thus, as Lemoine (1972b) shows, cancellation of the monstrous results of the incestuous union (by cutting up the flesh) gives rise to the origins of social differentiation between exogamous clans based on the incest taboo. The incest taboo is, therefore, the basis of social organization, and the theme of male sibling rivalry which we find in the later geomantic stories is already present in the myths about the flood, in the form of the mysterious elder brother who perishes. We can, I think, agree that all these stories form part of a single 'set' accounting for social and ethnic differentiation, from which the opposition between perishable and imperishable substances, associated with the opposition between an elder and a younger brother, as well as the opposition between Hmong and Chinese, emerges in a remarkably consistent way.

This was confirmed by another story which I received in the village, which also accounted for the loss of Hmong territory and mastery over the land as a result of Chinese deception. According to this account, which I will not for reasons of space reproduce here, the Hmong and Chinese had fought for mastery over the land for many, many years, but neither side was able to win. Therefore, they agreed to divide the land between them, by marking out separate areas of territory for themselves within it. While the Hmong demarcated their territory with *leaves and twigs*, the Chinese demarcated their territory with *stone* walls. Thus, the boundaries of the Hmong territory were swept away in the winds, while those of the Chinese remained forever (cf. Graham 1937).

As we shall see in the following chapter, such oppositions between wood and stone still characterize actual differences of Hmong and Chinese burial forms.

In a poetic way, such oppositions between perishable and imperishable materials, associated with the opposition between Hmong and Chinese, express the differences between the permanent-field economy of Chinese civilization, which has led to the establishment of great states and dynasties, characterized by power and the possession of writing, and the shifting, transitory culture of the Hmong, based on swidden agriculture and the ultimogeniture which is so often associated with it. It would be unwise to carry these oppositions too far, however, since the Hmong, too, know the principles of iron-working and use metal tools to cut down the forest, while their houses are not invariably of wood in China.

One is certainly tempted to see in the striking opposition between inauspicious metal and other materials, however, memories of the introduction of iron into China, in around 600 BC, which may have emanated from countries to the south of China, and enabled the earliest Chinese states to assert their hegemony over neighbouring peoples who then became the members of minorities, peripheralized into the deep recesses of the forests and the mountains (J. Needham 1965, p. 93).

Since this is not intended to be a historical study, I will not pursue these questions further here. But I do believe that in the sort of myths and legends I have presented, we find encapsulated many memories of material historical conditions, of which the opposition between perishable and imperishable substances is only one. Others would include Imperial desecrations of rebel pretenders' ancestral graves, the use by the Chinese of minorities in their wars, the revolts of the Miao against the Chinese, the dispossession of their lands and their marginalization, the failure to acquire literacy, and the exodus out of China. These *dab neeg* do, through their oral manner of transmission, show an extraordinary capacity to express and encapsulate changing social conditions and circumstances in terms of traditional symbols and images.

Moreover, these oral legends tell us a great deal about the Hmong sense of their own ethnicity, and how it is intrinsically opposed to that of the *Suav*, or *Mab-Suav*. It is the continuing transmission of these accounts in the village today which shows us part of the ongoing power and vitality of that conceptual framework (where 'here' is classical China, and 'now' refers to a timeless past), which *frames* the village in terms of the situation of the Hmong in Imperial China, and, above all, depicts the opposition between 'Hmong' and 'Thai' as, ultimately, one between 'Hmong' and 'Suav'.

---

1. Feuchtwang (1974, p. 221) suggests that geomancy may be 'used to express both difficulty in accommodating change and *also* the fact of having accommodated it'. While all the cases of its *application* among the Hmong confirmed only the former, it may be possible to view the legendary formulations of geomancy considered later in this chapter as performing something of the *latter* functions.

2. Sometimes the *memtoj* are identified with *cag dej*, the roots or sources of the *waters*, in distinction from the *loojmem*, identified as *cag roob*, or the roots of the *mountains*. Very often, however, the two are used interchangeably.

3. The White Tiger and the Azure Dragon represent the two fundamental polar oppositions of the system of geomancy practised by the Chinese.

4. See the myth of the degenerating ages attributed to Hesiod in Dodds (1928). Dodds (1973) describes how the two myths of the lost paradise and of eternal recurrence militated against the emergent notion of linear progress among the Greeks (cf. Nisbet 1980). See also discussions on the nature of time contained in messianic formulations in Burridge (1971, p. 148) and Laurence (1982).

5. The phrase 'when I am 120 years old' is a synonym for death, since death is not usually discussed, and even geomancy had to be discussed outside the house, and preferably after dark, for this reason. Geomancy is most often discussed, for obvious reasons, at funerals.

6. In the version I received, after Xob (Thunder) had struck, and the Great Dragon had caused the waters to cover the earth, all survived but *four* brothers and sisters. One couple gave birth to a gourd, the other to a gourd-shaped piece of flesh. From the divisions of the flesh came all the people with clans; from those of the gourd, cut with iron, came all those without clans (first the black ones, then the white ones).

9

'Real History' and the Theory of Ethnic Categories

THE most remarkable feature about the two main stories we have considered in the previous two chapters, that is, the legend of the two brothers contesting for *loojmem*, and the legend of Tswb Tchoj, is that they should so clearly associate Hmong allegories of messianic leadership and rebellion against the Chinese, with a geomantic system which has been seen as peculiarly Chinese. Thus, the entirety of the tale of Tswb Tchoj is couched in terms of traditional geomantic symbols, yet no sense is conveyed (nor is any in the story describing the fission of the Hmong from the Chinese in terms of the same system) that geomancy is not at least as Hmong as it is Chinese. By implication, the theme of the two brothers whose rivalry for an auspicious future is phrased in terms of geomancy, refers to a primordial stage of ethnicity in which the opposed categories of 'Hmong' and 'Chinese' had not as yet become clearly separated. It was *on account of the rivalry* between two brothers, one older and one younger, itself phrased in terms of a system for the burial of patrilineal ancestors which the Hmong and the Chinese *both practise*, that the divergence of Hmong from Chinese is said to have occurred. And it is at this point that we must ask what the ethnic labels 'Hmong' and 'Chinese' really mean, and also what is the historical status of the kind of stories we have been considering. In order to do this, it will be necessary to examine more closely in this chapter some details of the kinship system of the Hmong.

For it is a particular feature of the Hmong kinship system, first, that many of their surnames should be identical with those of the Han Chinese, leading Ruey Yih-fu (1960), for example, to infer that the patrilineal and patrilocal social structure of the Miao must have been borrowed from that of the Chinese at an early stage (probably after the Sung dynasty),[1] and, secondly, that Hmong clans should be divided into 'Hmong' or 'Chinese' (*Suav*), according to the type of burial practices followed by different descent groups, and the siting of their ancestral graves.

Thus, I might merely describe myself as a 'Hmong Lis', that is, a Hmong of the Lis clan or surname, but if pressed, particularly by another Hmong Lis, who might be trying to establish how nearly we were related, I would then have to specify whether I were a 'Hmong Lis Hmong', or a 'Hmong Lis Suav' (that is, a Hmong Lis of the *Hmong* kind, or a Hmong Lis of the

Chinese kind).[2] Now, although other distinctions of a ritual kind are also involved in formulating this dichotomy,[3] the distinction between 'Hmong' and 'Chinese' types of descent groups (*dab qhuas*) *among* the Hmong refers most importantly to differences of burial siting, in that the Chinese type necessitates burial of the deceased laterally *across* the line of the mountain ridge, while that of the Hmong type necessitates burial of the deceased horizontally *along* the line of the mountain ridge. But further differences are also involved, for the 'Hmong' or 'really Hmong' type of grave is simply marked with branches and brush, while that of the 'Chinese' type is protected by a mound of stones upon the grave. There is also an intermediary type of grave, which distinguishes other descent groups, which is fenced.[4] We have already seen, in Chapter 8, something of the part the distinction between permanent and impermanent substances, such as wood and stone, play in legendary formulations of the ethnic differences between the 'Hmong' and 'Chinese'. But here we see how the opposition is typified in the actual burial practices which distinguish descent groups among the Hmong—and distinguishes them, moreover, in terms of whether they are 'Hmong' or 'Chinese'.

Commonsensical explanations for these burial practices are that they keep animals away from corpses, or 'big people like trees who live in the forest', or grave-robbers, since there was a time in the past when people were so poor that they had to rob graves for the clothes which are buried with corpses.[5] Another explanation for the 'Chinese' type of grave was that it was done to prevent the *Chinese* from desecrating Hmong graves as they had done in the past (cf. Lee 1981, p. 296), and thus it became known as a 'Chinese' type of grave. However, as we shall see, this represents a typical internalization of oppositions which were originally exterior.

Other minor variations mark the differences between descent groups, such as how many times the grave is encircled at the burial, and dietary prohibitions, such as that on eating the hearts of animals which distinguishes segments of the Yaj clan, or that on eating their spleen which demarcates subdivisions of the Lis clan. It may be that, following the crossing from China, under conditions of famine and depopulation, these descent groups represented emergent sub-clans between which intermarriage was for a time permitted, as this did, in fact, happen with the Yao (Lemoine 1982a, p. 17). What concerns us here, however, is why such divisions among the Hmong should be termed 'Hmong' or 'Suav', and why they should be associated with burial practice.

It is a fact that intermarriage between the members of different ethnic groups in the region is very common, as is also the adoption of the children of other ethnic groups. The most obvious cases in the survey site were, in the focal village, the Khmu' who had married the niece of the old man of the village, who had to all intents and purposes 'become' Hmong, and whose children were expected to marry, and remain, as Hmong: in a neighbouring settlement, a Hmong woman who had married a Chinese shop-keeper; in the larger settlement, the Hmong woman who lived with her Khonmuang husband on the outskirts of the village; and the non-Muslim Chinese shop-keeper, who had married a fourteen-year-old Karen girl. When this marriage broke up, after the disasters described in Chapter 2, he had begun negotiations to marry a Green Hmong girl who lived outside the survey site. Adoptions, too, were not uncommon; I knew of several cases in the larger settlement, but of only one small boy who actually knew that he had been bought from his Karen parents, and once told me sadly that he had no *kwvtij* (clan relatives), since such children are usually brought up as Hmong in every way, with no apparent distinction between them and any other member of the community. Certainly, although the Hmong are often regarded as an isolated and autonomous group, this sort of thing has been going on for a long time on a small scale and receives some traditional sanction in order, both in the case of intermarriage and in that of adoption, to compensate for infertility and repair the labour force. Among some other peoples of the region, such mechanisms of incorporation are even more endemic. Lemoine (1983) recalls visiting a Yao village where a *majority* of the male adults had been adopted in infancy from the membership of other ethnic groups, since the Yao suffer from particularly low rates of fertility (Miles 1973).

In my interviews on the *origins* of clans, which I collected from outside the survey site as well since only five (the Thoj, Xyooj, Lis, Vaj, and Yaj) were represented within it, I came across several which were said to have been recently formed, within the previous sixty years, from the marriage of Hmong women with Chinese who had settled in Hmong villages, as those in the survey site had done. The well-established Lauj clan, for example, traced their descent to a Chinese father who had married a Hmong wife because he was unable to have children. This seems to have represented customary practice, since the Muas clan were said (by a Lis informant) to have very similar customs to those of the Lis, after a Muas member had married a Lis girl at a time when there were no surviving male members of the Lis clan, which was revived from this union. In the same way, two Catholic Hmong boys studying in Chiangmai came from a small Xem clan, formed fifty years previously after some Chinese had come to a Hmong village and married Hmong women.

In contrast to Geddes' opinion (1976, p. 55) that commentators have been misled about the number of Hmong clans in Thailand, it is this which accounts for the fact that there are more clans than the ideal number of twelve referred to in ritual discourse (at least fifteen among the White Hmong). The Hmong Vaj, Lis, and Tsab, were claimed to be the most ancient of the clans, since they were the most numerous, while the Tsheej, Xem, and Faj were considered to be the newest, and the smallest.

As we have seen in the previous chapter, one of the most common beliefs in accounting for the *origins* of the Hmong/Chinese distinctions within Hmong clans themselves, is that they arose from the marriage of a single male, first to a Hmong, and secondly to a Chinese woman—thus from two half-brothers (classed with full brothers in Hmong kinship terminology). To illustrate this point, I will extract the following from a much longer account by Nyiaj Tub Yaj (Loei, March 1982) of the story of Creation which expresses this particularly clearly:

The First and Second Wives

It was at that time (after the Flood) that the Hmong and the Chinese were fighting together. Because they had a King, they were able to vanquish the Chinese. At that time China had ten cities. The Hmong King had a most beautiful wife named Nkauj Ntsuab, and his name was Nraug Nas.[6] When the Chinese came and saw how beautiful Nkauj Ntsuab was, he told Nraug Nas how beautiful she was, and suggested that they exchange their wives. And Nraug Nas agreed. But then the Chinese people asked their King why he should have to give his own wife in exchange for Nkauj Ntsuab. 'If you want her, just take her,' they suggested. 'Nraug Nas can find a new wife.'

And the Chinese King agreed. So he came and took Nkauj Ntsuab away, but darkness fell upon the earth for seven years. All that time Nkauj Ntsuab and Nraug Nas were searching for each other. One day Nkauj Ntsuab came to the great mountain at the foot of heaven, and there she found Nraug Nas. Nraug Nas was crying because he had been unable to find her. 'Why are you crying?' she said. 'Don't you remember that, at the beginning of time, our ancestors said that if ever anyone should come to take either of us away for their husband or their wife, you must be sure not to forget your *qeej plhaub mag* (reed pipes of hempen skin). When you play it, we shall meet again.' And he answered that he did remember. At that time the Hmong knew many things. They knew the uses of magic.

So Nraug Nas took up his *qeej* and played it, and Nkauj Ntsuab laughed three times, and then all the earth was bright again. Then the Chinese King said, 'Nraug Nas, ever since I took your wife, the earth has been dark, but now it is light again. Why don't we exchange clothes, you put on mine and I will put on yours, and you play the *qeej* again, and let Nkauj Ntsuab laugh again, and the world will become still brighter.' So they exchanged clothes, but as Nraug Nas played the *qeej* it broke, and so did Nkauj Ntsuab's laughter, and rocks rained down from the heavenly mountain upon the Chinese King who was standing there, and he died. They threw his body into a great cave. But three years later, the Chinese King rose up again, and made war on Nraug Nas, who lost the battle and fled, and was caught by the Chinese. According to the elders, it was 1700 when the Chinese made war with the Hmong and caught Nraug Nas. And that was the time when the Hmong first came over on to this side. Later they had another King, named Yaj Yuam. The Chinese were winning at first, but then the Hmong captured nine cities. When the Chinese saw they were about to lose, their King gave his daughter in marriage to Yaj Yuam. And that is why we sometimes ask each other if we come from the Hmong mother or the Chinese mother.[7] Although we have never been back to see, some say that the Hmong King's grave is still there in China.

Here, in legendary form, the presence of Chinese subdivisions within Hmong clans is directly attributed to intermarriage with the hostile Chinese. As Chapter 8 notes, it is often said that the Hmong and Chinese *sib ncaim* (separated, divorced).[8] Although it is not surprising that dietary prohibitions and other minor variations should distinguish Hmong descent groups, it is remarkable that ethnic differences between the 'Hmong' and 'Chinese' should also be involved in distinguishing them in this way.

We see here, that it is above all the *patrilineal naming system* of the Hmong, through the burial practices which define it, which is associated with the *geomantic system* which orders the siting of such burials.

Kinship, burial, and geomantic practices all form part of a *single conceptual system*, characterized by powerfully 'Chinese' overtones. However, this conceptual system is not unique to either the Hmong or the Chinese, but seems to be *shared* with many other border minority peoples.

My first intimations of this came when certain Hmong informants claimed that they shared kinship categories, and thus could intermarry, with the Lisu (as they also do with the Yao), a Tibeto-Burman group who inhabited similar regions of south-western China to the Hmong in the past, and today predominate in Burma and Northern Thailand. Further research showed that it is indeed the case that the Lisu also employ Chinese-type surnames like the Hmong. Furthermore, Lisu lineages are divided into 'really Lisu' (Lisu *tzy tzy*, who never eat new corn) and 'Chinese Lisu' (*hepa* Lisu, who do eat new corn), just as a Hmong may describe himself as 'Hmoob *tiag tiag*' (really Hmong) or 'Hmoob Suav' (Chinese Hmong). Thus, some groups of 'really Lisu' Lisu are not supposed to marry into a 'Chinese' group of Lisu, and fines were demanded when such liaisons took place (Durrenberger 1970). Similarly to the Hmong, the presence of Lisu lineages with Chinese names is explained by the Lisu themselves as resulting from intermarriage with Chinese: 'Most Lisu prefer to be "Chinese" Lisu, or Lisu whose ancestors were Chinese', and 'Most Lisu lineages are not uniquely Lisu but are shared by the Chinese at least and *probably by some groups of Lahu*' (Durrenberger 1970, p. 252, author's emphasis).

It should be noted that the Lahu, another Tibeto-Burman group of the same general region, are usually accorded the status of an ethnic group in their own right. Although their kinship is of a cognatic kind, Durrenberger (1970) reports the following conversation with a Lisu:

'Are you a Lisu?'

'Yes.'

'What group do you belong to?'

'Lahu.'

And he adds that one Lisu clan name, the Sha Kai Shi, is in fact the name of 'a kind of Lahu' (Durrenberger 1970, p. 252).

To explore this a little further, Leach (1954, p. 45), whose work was largely concerned with explaining how it was possible for the Kachin to 'become' Shan, shows how the *Lisu* were classed as one of the kinds of the Kachin people, while Hill (1982) documents cases of the *Shan* people in the Tengchong area of Burma adopting Chinese names from the sixteenth century onwards. But this sort of ethnic ambivalence and confusion in the region is not at all uncommon. To give just two further examples; Kunstadter (1979) recounts a similar conversation to the one above, with a girl who first claimed to be Lua', but then admitted that she was a Karen 'with Lua' relatives', and Hanks (1965) documented the emergence of what he termed a 'new ethnic group': the Lahu Shi Hopoe. This group had descended from the marriage in the first generation of a Shan to an Akha woman, whose children had married Lahu. Their daughters had married Chinese who lived among the Lisu, whose children in the fourth generation had taken Lahu spouses again. It does seem certain that the patrilineal naming system of the Hmong was not only not unique to the Hmong, nor shared by them only with the Chinese, but rather formed part of a common system with other peoples in the past. It is for this reason that we began by suggesting that Hmong social organization be considered a 'sub-nuclear' type of social system after Lehman's (1963) model of Chin relationships with the Burmans. I would hypothesize from this that the geomantic system practised by the Hmong, with its clear associations with the burials of patrilineally reckoned ancestors, similarly formed a part of a shared and inter-ethnic system in the past: a *common conceptual system* which defined the village inserted in this case into the structures of the Thai state.

Most ethnographers of the region have chronicled the chronic ambivalence of ethnic categories there, and the fluidity with which ethnic affiliations may be changed (Leach 1954; Lehman 1963, 1967a; Moerman 1965; Naroll 1968; Durrenberger 1970; Hanks 1965; Hill 1982, among others). The region was one of constant, shifting changing of ethnic boundaries, their memberships and markers. Leach declared that 'any particular individual can be thought of as having a status position in several different social systems at the same time' (Leach 1954, p. 286). Lehman claimed with regard to the whole region, that entire communities might be faced at any time with a *conscious* choice about which ethnic group to belong to (Lehman 1967a).

The gradual Sinicization of the southern regions of China took place under a system of local administration which replaced earlier tributary relationships the Chinese had maintained with the rulers of neighbouring states, from the time of the Sung dynasty onwards. Under this system of 'using barbarians to control barbarians': (Hill 1982), local leaders were appointed by the Chinese court as *tu-si* (local chieftain) officials over their own peoples, in exchange for offerings of annual tribute and the adoption of patrilineal, hereditary surnames. Some Miao certainly became *tu-si* officials (the Viceroy of Yunnan in the latter part of the nineteenth century was a Miao, as was the military governor of Yunnan in the 1930s: Fitzgerald 1935, p. 8). But it is more likely that Miao would have been associated with Lolo landlords who more frequently became semi-independent *tu-si* officials. De Beauclair (1956a) records how the Miao of North-west Guizhou would accompany their Lolo *tu-si* to the court of the first Emperor of the Ming dynasty (1368–99), and that this was still remembered by their descendants. Certainly patrilineal surnames could have been adopted by the ancestors of the Hmong through their association with Lolo landlords who had become *tu-si* officials and adopted such names instead of their own system of linked patronymics. As Hill (1982, p. 79) points out, the process of Sinicization was never a one-way process:

The 'Sinicisation' process noted in several sources was not so much the result of an invasion of south-western China by Han Chinese from the North as it was the result of indigenous peoples, especially in lowland areas, becoming Chinese. Thus, many of the 'Chinese' in this area were not descendants, in the biological sense, of groups of northern Han-Chinese; rather, they had adopted the Chinese role when it had become advantageous to do so.

Schafer observes, too, that most of the population of southern China were not really 'Chinese' at all, and prefers to call them 'creoles' throughout (Schafer 1967). Hsu's ethnography of a 'Chinese' community (Hsu 1949) provides a further instance of this (cf. Li Chi 1928).

At the same time, however, as members of indigenous minorities 'became' Chinese, so too those ethnic Han Chinese who did migrate to the south from the north very often, as we have seen, married members of those indigenous minorities and settled among them, becoming members of those minorities themselves. In Guizhou, the descendants of former military colonists who had married tribal peoples were known as *lao han ren* to distinguish them from the new (guest) settlers, who were called *Hakha* (De Beauclair 1956b). Lin Yueh-hua (1940) cites Chinese sources for evidence that, as early as the period of the Five Dynasties (907–60), soldiers garrisoned among the Miao had settled down among them and formed subdivisions within the Miao. Certainly many Muslim soldiers remained garrisoned in the south after Kublai Khan's conquest of Dali (Yunnan) in 1253, and it is the Muslim Chinese with whom the hillpeople most frequently came into contact.

I would present the hypothesis that, since we have no means of telling whether the actual ancestors of present-day 'Hmong' identified themselves as Hmong, Chinese, or the members of any other group, and no means of telling to what ethnic groups the ancestors of the present-day 'Han Chinese' in southern China belonged, there are no ethnographic or historical reasons for assuming that geomancy, or the patrilineal naming system with which it was associated, was in its origins or is now a distinctively 'Chinese' system at all—since the definition of what constitutes the ethnic identity of the 'Chinese' itself seems to be in considerable doubt.

If this hypothesis is accepted, then the mystery of why the Hmong should employ the geomantic system to express their own ethnic oppositions to the Chinese, disappears. The hypothesis, then, is that geomancy, most prevalent in the southern regions of China where the most complex processes of inter-ethnic assimilation took place, as well as many other aspects of the patrilineal naming system and burial practices normally considered 'Chinese' (such as the practice of double burial), must, like the ethnic distinctions themselves between those people who today identify themselves as 'Hmong' or 'Miao' and those who identify themselves as 'Han Chinese', have been formed out of complex historical processes of integration, incorporation, and assimilation associated with the formation of the Chinese state. Indeed, as Feuchtwang (1974, p. 121) points out, the geomantic system is above all associated with mountains, although practised in many areas of China where no mountains existed, and it would seem reasonable to assume that the system originated in the mountainous regions which have always been associated with the ancestors of the Hmong.[9]

We must ask, then, what kept the two ethnic terms, 'Hmong' and 'Suav' in Hmong formulations, 'Miao' and 'Han ren' in Chinese ones, apart. Here Lehman (1979) has introduced a most valuable distinction between 'an ethnic category and the genetic–linguistic groups with which it may at times be identified'. Thus, in Northern Thailand, the ethnic label 'yang' does have a historical constancy in referring to a given type of people inhabiting a particular ecological zone, whether or not the people to whom it was formerly applied were the ancestors of the present-day Karen to whom it is applied.

For the case is precisely similar with the ethnic category 'Miao' used by the Chinese for more than 2,000 years to refer to a barbaric, rebellious people on the outskirts of the Chinese Empire, and still used today to distinguish the Hmong in southern China, whence the current Thai derivative for the Hmong, 'Meo'. As Ruey Yih-fu (1962) shows, the term 'San-Miao', used in the works of Mencius, the *Shu Ching*, and the *Ta-Tai Li Chi* to refer to a people banished by the legendary Emperor Shun (2255–2206 BC?) to San-wei, disappeared after the Ch'in period, apart from a brief mention in the *Man Shu* of the 860s, and did not reappear as the name of a specific ethnic group until the 1190s. Ruey thus divides the periods of the Miao into a legendary one (*c.*2300–200 BC), an unidentified period in which the Miao were not distinguished from the generic term for barbarians, 'Man' (*c.*200 BC–AD 1200), and the modern period (AD 1200 to the present).[10] The question is, of course, whether the modern Miao, of whom good records exist since the Sung dynasty, bore any relation to the ancient Miao who, according to Ruey (1962), 'were inhabiting the midwest of China proper in the last quarter of the third millenium BC'.[11]

This is not a question, as it might seem, to be simply dismissed outright, for as most commentators on their culture and a majority of great Sinologists (Granet 1930; Maspero 1950; Eberhardt 1968; J. Needham 1965) have noted, the present-day 'Miao' retain strong traces of the most archaic forms of Chinese social organization known. Examples of these are the courting ball games still played by the Hmong at the New Year, and the bronze drums used by some Miao of Guizhou (De Beauclair 1960). It is in the context of such questions as these that the kind of *historical consciousness* expressed in the legends and oral traditions we have considered becomes most suggestive. Many authorities, moreover, do consider that the post-Sung dynasty Miao are identifiable with the ancient 'San-Miao'.

However, if one adopts Lehman's (1979) distinction between ethnic categories and the genetic–linguistic groups they refer to, such questions become, quite simply, unimportant. What is of primordial importance is that the same term (Miao) has been employed to distinguish a rebellious and unsubjugable people over some 3,000 years of recorded history, from the majority population of the Han Chinese. And this is precisely what the stories we have been considering are telling us.

For example, the Tswb Tchoj allegory depicts a Hmong and a Chinese representative vying with each other for control over a specific area of territory. This sort of contest constitutes the essence of the geomantic system which, in the account of the two brothers (Chapter 8), is blamed for the separation of Hmong and Chinese identities. Whether or not one believes, as Wiens (1954) does, that the Miao were the original, aboriginal inhabitants of China from whom the modern Hmong descend, forced further and further southwards by the encroaching march of the Han Chinese invaders, this is certainly the kind of picture which the legends, from the legend of the loss of writing to the legend of the defeated Hmong Emperor, are presenting. And in particular, we do know that fierce historical conflicts have taken place throughout recorded history between the 'Miao' and the 'Chinese'.

Wiens (1954) chronicles more than forty Miao uprisings between AD 403 and AD 561. During the Five Dynasties period, Ma Yin, of the new state of Ch'u, subjugated the Miao. During the Ming and Qing dynasties, there was hardly any time when suppression or pacification campaigns were not being undertaken against the Miao (Wiens 1954, p. 135). Clark (1894) lists Miao-tse uprisings between 1630 and 1730. Lombard-Salmon (1972) describes the great Miao rebellions of the eighteenth century. The last great rebellion of the Miao in Guizhou took place, as we have seen, in 1856, and required some 20,000 Imperial troops to subdue it, while we have already given some account of rebellions which have taken place throughout the present century. Against this sort of background, the myths and legends considered in this book provide a very clear record of the cultural *impact* of what actually occurred.[12] However, it is not what actually happened which is the main concern of this study. What is, here, the main concern, is how what actually happened is *experienced* today, and to what extent such 'recollections' of the past influence present and future behaviour. What is important is how the present-day Hmong *experience* their past history, and explain their *current* situation of the deprivation of power, literacy, and territory. And this situation is itself, as we have seen, a historical 'fact'.

History itself seems, from this point of view, more and more to resemble what we mean by ethnicity, when we say that it is very much a matter of *conscious* choice (Saul 1979; A. Cohen 1969; Lehman 1967a; Banton 1977). We select our own histories, which are the significant events for us *now*, isolated from the mass of events which we have truly encountered, and they become 'real' to us. It is the constitution of a *significant*, or a *real*, history, which I am in search of.

For the Hmong, it seems, the *significant* parts of their more recent history have been: their original claim to the lands of China; the deprivations and depradations of the Chinese; their encounters and conflicts with the Chinese; the victory of the Chinese; the flight or exodus of the Hmong out of China through water; the loss of their power, territory, and writing; and their settlement in their new lands of abode. In the following chapter, we see how similar myths are beginning to emerge which depict the latest and most traumatic historical conjuncture for some groups of Hmong: their defeat by the Pathet Lao and dispersion to homes scattered throughout the world. But first the concept of a real history should be explored further, since we have seen that throughout the oral tradition of the Hmong, the Hmong historical past is phrased by them in terms of the alternatives between being Hmong and being Chinese—just as their current present in Thailand is phrased in terms of being Hmong or being Thai, while the future is being increasingly phrased in terms of being Hmong, or becoming French, American, etc.

Just as there is a sense in which we may speak of the twenty-first century, and say that, given certain conditions, there may well *be* no twenty-first century, so it is evident that history as we understand it depends upon our participation in it; that is, there can be no history without a historical consciousness of it. *It is the status of this kind of historical consciousness which I am seeking to establish.*

It is possible to be still more specific about the origins and ending of 'history' than the Hegelian formulation of it as the progress of consciousness in liberty would allow. Thus, Feuchtwang (1975) draws attention to Althusser's (1971) argument that 'we are to understand by history itself a limited concept; ideology is limited to those societies with history, which is characterised by class struggle and a state apparatus' (Feuchtwang 1975). In Chapter 8, we considered how the phenomenon of 'hypostasis' manifested in the periodic recurrences of appeals to the 'legend' of Tswb Tchoj illustrated Levi-Strauss's assumption that there are 'peoples without history', that is, that there are 'cold' societies which, 'although situated in history . . . seem to have developed or retained a particular wisdom which impels them to resist desperately any modification in their structure which would enable history to burst into their midst' (Levi-Strauss 1977). If Levi-Strauss is right, these are particularly those societies which lack a form of writing, and are possessed of an *oral* tradition for the transmission of knowledge—societies which have not attained that degree of self-consciousness brought about through the interaction with other ethnic groups which would have enabled them to formulate a sense of history specific to themselves—and without which, as we have seen, there can be no history, in the proper sense of the term. From this point of view, the Hmong would represent an intermediate case, since they demonstrate such a powerful awareness of their lack of writing, and have moreover, as we have tried to show at several points in the course of this book, maintained interactions with the members of other ethnic groups for a very considerable length of time. Ross (1979) has distinguished the communal, minority, ethnic, and national groups according to the extent to which each becomes conscious of itself, its unity and differences from other groups, through the encounter with members of other cultures.

Here we are at the heart of the hidden relationship which exists between ethnicity and history, for as Lehman (1979) points out, 'It is too little understood that cultural change amounts precisely to . . . a change in ethnicity.' This is a large, but a logical, claim, which has to be taken into account in any consideration of a 'transitional' community of the type represented in the survey site. Thus, in considering ethnicity, its margins, and variations, we are considering the process of historical choice itself.

The only alternative explanation to assuming that (as their legends tell us) the Hmong really *did* once have an Emperor, a country, and a form of writing of their own, is to posit that, in order to explicate their own ethnic differences from the members of other majority and more dominant cultures, the Hmong have been forced more and more to *define their own ethnicity in terms of what it is not*, hence, *absence* of writing, *absence* of territory, *absence* of kings and emperors. Thus it may be that current ethnic deprivations have been projected into a distant past in order to justify their own, felt, ethnic distinction from the Chinese, who *were* characterized by all those things. This explanation does not, however, invalidate the concept of real history, since these deprivations must themselves have had historical causes, and it is these which are spelled out in the myths and legends accounting for such (negative) characteristics. Thus, the more I define myself by saying I am *unlike Mr B*, the more I come to define myself *in terms of Mr B*, and what Mr B is. In this case, a general conclusion would be that, the more a minority culture seeks to define itself in terms of its own opposition to a majority culture, the more it is forced to appropriate the very terms and symbols employed by that majority culture.[13]

A very common distinction employed by the Chinese to characterize different types of minority populations according to their conceptual distance from the Chinese state was that between *sheng* (raw) and *shu* (ripe). This dichotomy distinguished Miao groups from at least the mid-sixteenth century onwards (Ruey Yih-fu 1962). 'Shu Miao' would have referred to people who had accepted Chinese suzerainty, spoke some Chinese, had perhaps adopted some form of Chinese dress and names, and lived nearer the valleys practising permanent-field agriculture. 'Sheng Miao' would have referred to groups of the same people who had remained up in the high mountains, remote from the Chinese state and at times attacking border posts, practising swidden or pastoral forms of agriculture. It may have been this widespread dichotomy which led later ethnographers to distinguish, for example, between the 'Wild Wa' and the 'Tame Wa' (Endriquez 1924; Stevenson 1944) or even between 'gumsa' and 'gumlao' Kachin (Leach 1954). The most remarkable feature of the dichotomy, however, was that it was originally not applied to distinguish different groups of minority peoples from one another at all. Originally, it was used to differentiate the Han Chinese settlers themselves, seen as ripe, mature, and civilized, from all other minority peoples, seen as raw and uncivilized (Hill 1982, p. 76).

The *internalization* of such an opposition, although still the result of external formulations, does bear a remarkable resemblance to the way in which the ethnic distinction between 'Hmong' and 'Suav' has itself been internalized, this time by the Hmong themselves, within the kinship categories of Hmong society. In both cases, an original opposition between a majority and a minority culture, has been internalized within the culture of the minority.

Certainly, far-reaching contradictions must have been introduced into the social organization of the 'Miao' peoples through their contacts with the Han Chinese, and the long historical process of assimilation which took place on both sides, and this may explain to some extent the presence of Hmong- and Chinese-type burials, graves and descent groups *within* the Hmong social system. But again, one has to beware of over-reifying ethnic categories in the first place. Moreover, those people who chose to remain as *independent* 'Miao' as other groups of 'Miao' fell under 'Chinese' influence, cannot have been unaffected by this process, and to an important degree, would not have been 'Miao' in the same sense in which they would previously have been 'Miao', before their culture had been divided by contact with the members of more dominant groups. We may express this as follows: 'Where A stands for "Miao" and B stands for "Chinese", A *versus* B bifurcates to form AA and AB.'

Here, while the contact with a dominant culture results in the internalization of the opposition between that culture and the minority culture in one aspect of the minority culture (AB), at the same time another aspect of the ethnicity of that minority culture is strengthened, intensified, and reinforced through its opposition to the dominant culture (AA). And it is in the realm of AA that we may seek the origins of the emergence of messianic movements among the Hmong (provided that this explanation of what actually happened is accepted).

In this light, the Tswb Tchoj story can be seen as the symbolic paradigm which resolves the fundamental problem of Hmong ethnicity: how to be not Chinese, and Chinese.

The logical impossibility of Hmong sovereignty over China, and of a Hmong Emperor of China, is depicted in the inhuman descent of the hero, Tswb Tchoj, from the union between a woman and boar. Thus the myth is essentially tragic in structure, and it is with the knowledge of certain defeat and a sense of despair that it is activated in messianic form at moments of crisis. And it is in this sense that such messianic movements are truly heroic. In a very similar way, the myth of the descent of the Yao clans from the union between a Chinese princess and a magical dog (P'an Hu) depicts an 'impossible achievement' (Levi-Strauss 1963b). But here the myth is activated, as Lemoine (1982a) shows, to legitimate the traditional exemption of the Yao from taxation and corvée labour, and their freedom to practise shifting cultivation.[14] For the Yao clans, unlike the Hmong Emperor of China, actually exist. R. Davis (1974) has drawn attention to the 'surprising functionalism' of Levi-Strauss's general thesis that, through demonstrating the untenability of alternatives, myth may validate actual social practice. In the legend of Tswb Tchoj, we see how this is so. We see, also, how it is possible for the messianic movements among the Hmong to represent a nativistic reaction to the schisms brought about by the internalization of Chinese categories of discourse. If the Hmong/Chinese distinctions we have considered within Hmong social organization at the level of burial practice and kinship usage represent a *centrifugal* tendency in which Hmong society is brought into contact with that of the Chinese, then the powerful recurrences of messianism and revolution in Hmong history may be seen as representing the *centripetal* force which seeks to *counteract* these fissive tendencies, and bring about a new unity for the Hmong, by returning to the sources of Hmong ethnicity.

While this provides, in my view, a plausible explanation of the mechanics of the sort of adoptions of the Hmong from the Chinese which may have occurred, one can never forget that in dealing with the ethnic categories of 'Chinese', 'Hmong', 'Miao', 'Lolo', and others, one is not necessarily dealing with consistent genetic–linguistic groupings which have remained isomorphic with their ethnic categorizations over historical time. It is for the purpose of overcoming this sort of dilemma that the notion of 'real history' has been introduced as a heuristic device, since it enables one to overcome the problem of dealing, in an anthropological context, with the questions of who borrowed what from whom over historical time. Through examining the Hmong practices of supposedly

'Chinese' system of descent, burial, and geomancy, and their use of geo-mantic ideas and symbols to describe their own messianic struggles against the Chinese, we are, in fact, examining what the majority of other com-mentators, *unless* they have assumed that the Hmong were the aboriginal inhabitants of China, have supposed to be the very extensive Sinitic in-fluences upon their culture. Through introducing the notion of real his-tory, on the grounds of the considerable flexibility of ethnic categories which previous ethnographers have shown to be characteristic of the region as a whole, we have tried to transcend this problem and show that, in the final analysis, it does not matter whether or not the Hmong have adopted their practice of geomancy and many other cultural elements from the Chinese. Ethnographically, what matters is how the Hmong define their own ethnicity with reference to their own sense of the past, in which there is no awareness that these elements may have been borrowed, and it is this sense of the past which this book has tried to present.

1. This view receives some confirmation in De Beauclair's (1960) citation of a mid-sixteenth-century account to the effect that the Miao at the time had no family names. How-ever, according to De Beauclair, the sources are contradictory, and one must remember that, historically, the relationship between the ethnic categories 'Hmong' and 'Miao' may not have been a consistent one.

2. It is probably this kind of thing which leads Geddes to conclude that the Hmong term for 'clan' is 'Hmong' (Geddes 1976, p. 55).

3. In particular, at the *dag roog* ritual performed ideally every three years for the spirit of the bedroom, and the *ua nyuj dab* ritual, the last of the mortuary rites, performed in the case of severe illness of a household member diagnosed as having been caused by the restless soul of an ancestor for whom a cow has not yet been sacrificed.

4. Another variant is to erect a rough shelter over the grave.

5. Similar practical explanations are given for the practice of burying male placenta under the centre-post, although Lemoine (1972b) has shown there are important ritual meanings involved. It may be simply said that the practice is to keep animals away.

6. The original couple.

7. Because the Hmong King would also have had a Hmong wife!

8. Although it did happen that the hands of Chinese princesses were offered to the chief-tains of neighbouring 'barbarian' peoples, such as the Hsiung-Nu, legendary Hmong formu-lations in which a Hmong King receives a Chinese princess more probably represent an inversion of the long historical process in which Chinese settlers in Hmong villages took Hmong wives, as we have seen occur within the survey site.

9. See Fried (1975) on the evolving centre–periphery oppositions which characterized the formation of the Chinese state itself. On the difficulties of ethnic identification in modern China, and particularly Yunnan, see Hsiao-Tung (1980) and Odayashi (1970).

10. White Miao were distinguished from the Blue and Flowery Miao as early as the Yuan dynasty (De Beauclair 1956b).

11. Shiratori (1966) has, for example, denied that the 'Miao' are a homogeneous group at all, but rather represent an amalgamation of different ethnic groups, taking the title of 'Miao' from a family name common among the Yao.

12. Binney (1971, p. 4) is quite wrong to declare that there are 'almost no historical records of the Miao'.

13. I have throughout followed Tafjel (n.d.) in using 'minority' to refer to a social status rather than a numerical power, although, in fact, in all cases I know of cited in this study, the two do coincide.

14. For other versions of the Yao ancestral myth, see Liu Chung-shee (1932); Lee Chee-Boon (1939); Yoshiro (1975), and Tan Chee Beng (1981).

The Reformation of Culture

THE centrality of the Tswb Tchoj tale to Hmong custom and tradition, and the power of the legend to deal with and articulate changing social circumstances and continually define Hmong ethnicity in terms of traditional symbols, is shown in the ending of the tale of Tswb Tchoj in America (Chapter 7). The raconteur, Xauv, was an old man, of the larger settlement, who had never visited the refugee camps nor had any contact with Laotian Hmong. Yet, as I have pointed out, members of the village had visited the refugee camps for reasons of their own and had brought back some news of the Hmong there and their reception by foreign countries, particularly America. Although Xauv possessed no radio of his own, he will have been familiar with the situation of the Hmong refugees from Laos through the Chiangmai and other radio station broadcasts on the transmitters of other villagers. As I began this book by remarking, the village studied was as much isolated as it was not isolated. While Hmong in the village had access to news about the Falklands War and had heard a grisly tale about a Hmong in Brazil who had sold one of his eyes and a kidney for money, at the same time I could be asked, by a bridegroom sitting outside the door of his wife's father's house in Hapo, whether there was a sun in America. And from such materials legends are created.

One receives additional evidence, moreover, of the significance of the paradigmatic symbol of Tswb Tchoj in the definition of Hmong ethnicity, in the extraordinary emergence of the new messianic movement in the refugee camps, which is focused on the memory of Tswb Tchoj in particular and has, as I have said, erected a large statue of Tswb Tchoj's father, the boar, in the centre of its temple compound, close to a seminary where the priests of this new religion teach the divinely revealed Hmong writing.

For this reason we shall examine, in this chapter, some material on the Hmong refugee population from Laos, moving from a consideration of their situation in the refugee camps within Thailand, and the religious movement which has emerged there, to an examination of some of the social problems they have encountered in their new homes overseas. Many of these problems reflect back on the problems analysed in the context of the village situation, while new myths and legends have emerged which illustrate the attempt to depict this most recent conjuncture of the Hmong historical experience in terms of the metaphors and symbols analysed throughout the course of this study.

The Hmong refugees from Laos who have fled to Thailand tend to see their flight from Laos after the Pathet Lao victory of 1975 as a second 'exodus', comparable to that first exodus out of China to the lands of South-East Asia. While many have been resettled in 'third countries' overseas, such as Canada, French Guyana, the United States, France, and Australia, at the time of study five major refugee camps remained in Thailand along the border with Laos housing Hmong refugees, many of whom were too old, or sick, or depressed to be resettled overseas.[1] Having effectively lost the guerrilla war against the Pathet Lao in which many of these had been involved, and with no fields to cultivate, the refugees had turned to new ways of formulating their past and future circumstances. I was able to visit and stay in these camps in the company of a band of Hmong males from the survey site, who were curious to see whether these Hmong 'brothers' from Laos were like them. One of the main concerns in the camp at the time I visited it was to bear as many children as possible, since only by repairing the loss of lives suffered in Laos could the Hmong hope to regain their land there. Most of the Hmong I interviewed were extremely worried about whether they should remain in Thailand, with the hope of regaining their land in Laos, or be resettled overseas, since sorry tales of unemployment, language, and medical problems encountered by their relatives in their new lives overseas had already reached them.

The Thai government would not permit these refugees to settle in Thailand, although contacts of an informal type between the Lao and Thai Hmong were already taking place, mostly involving marriage to refugee women whose families could expect only small bridewealths, or the opium trade, since many of the older Hmong refugees were still addicted to opium. Lemoine (1972c) had described how it was in the huge refugee camp in Laos (for, since at least 1965, Hmong within Laos had been in the position of refugees: Dommen 1971) of Long Tien, among a population of 30,000, that the messianic moment associated with the name of Yaj Soob Lwj had first arisen. Under the impact of evangelical Christian fervour in the Thai refugee camps, associated with charity agencies who, for example, insisted on prayers before imparting the English-language training the refugees would need for their new lives abroad, this movement had gained a new strength and many more supporters.

In marked contrast to the shamanic Daoist religion of the Yao, which is characterized by the use of special paintings on ritual occasions for the depiction of the gods (Lemoine 1982a), and unlike, too, the Theravada Buddhist religion of the Thai and Lao, which is characterized by the prominent display of statues of the Buddha and temple murals depicting scenes from the Buddha's lives, in the religion of the Hmong there is nothing resembling representational divine art. Nor are there communal rituals or places of worship, or grades of priest distinguished by their special clothing. Yet, here in a refugee camp, the Hmong had created all these things. Within a special compound and according to geomantic principles, a temple had been constructed out of several octagonal huts, which housed a collection of idols of Hmong deities and culture heroes, as well as a series of remarkable musical instruments, such as two-legged xylophones and triangular drums. A special grade of priests (*txam meb*) with a distinctive dark uniform conduct rituals of worship and healing for the followers of the movement at designated times; that is, in the early morning and evening, as in a Buddhist temple, and also on Sundays. The movement is led by a charismatic prophet named Lis Txais, who is said to have designed the musical instruments himself, and whose many original songs and talks are circulated widely in cassette form, preaching a stringent morality, the need for unity among the Hmong clans, and the real history of the Hmong. Indeed, it is denied that there is anything new in the movement at all; the priesthood, rituals, musical instruments, and statuary are all said to have been rediscovered, and to be those which the Hmong had had in an original home now believed to have been Mongolia.

Even the name 'Hmong' has received an original interpretation through this movement, since Lis Txais explains the word *Hmoob* to have been derived from *hmoov*, pronounced with a rising tone, which means 'destiny, luck, or fate'. And the phrase *peb hmoob*, which the Hmong often use when referring to themselves, since *peb* means 'we', Lis Txais also explains as originally having referred to three Hmong brothers, since *peb* can also mean 'three', who founded the Hmong people. The main function of the priests is to teach the sacred writing revealed to the prophet Yaj Soob Lwj in a dream to many of the camps' schoolchildren. Yaj Soob Lwj was in direct communication with Tswb Tchoj, whose black boar father is represented in the form of a statue above the temple. The movement represents a spiritual response to the loss of land and territory, once again clearly linked to messianic ideals of royalty and the possession of a special form of writing, and I would argue also a response to the impact of Christianity which has won many adherents among the Hmong in the refugee camps and overseas, since one belief circulating in the camp at the time I visited it was that a pair of every species of animal in the world should be slaughtered before a new dawn of peace, unity, and prosperity would come to pass, in a direct reversal of the Christian parallel. The movement is synthetic in influence, incorporating both Christian and Buddhist elements, and teaches sympathy and respect towards the followers of other faiths.

The following legend, told by Yaj Nyiaj Tub (Loei, March 1982), is one of many which has emerged from the context of the refugee camps, and is of particular importance, like the ending of the tale of Tswb Tchoj in America, in illustrating the capacity of the oral tradition to incorporate radically new social conditions in an explanatory way. Thus, the embroideries which refugees in the camp make for sale and send to their relatives resettled in the United States are here given a poetic precedent in the form of the embroidery given to the Hmong King by the wife he has banished overseas, which he must keep as a token of their relationship until he is reunited with their sons who will come back to help him: in the same way, Hmong refugees in camps in Thailand send their embroideries to relatives overseas who will one day come back to help them regain their lost territories in Laos. Also, since certain receiving countries refuse to recognize the validity of polygynous families and therefore divorces often have to take place before Hmong refugees can be resettled, this is echoed in the divorces the Hmong King's sons have to have before they can cross the ocean. And beyond what is directly given in the context of the legend, the King's three sons are said to represent three groups of Hmong of whom it had been prophesied that one should go north to become powerful and control the others, one should return home, and one should go west across the seas to become invisible so that only their voices could be heard. This prophecy is said to have been fulfilled by the Hmong refugees of today, since some have been repatriated to China in the north, some have returned to their homes in Laos, and some have gone westwards over the seas, sending only their voices back on cassette tape, as if they were ghosts.

The Story of Xeev Xais

A long time ago the Hmong lived in the mountainous lands of Mongolia. At that time the Hmong had kings and power. According to the elders, the Hmong even had an Empire for a while. And at that time the Hmong had an Emperor, who was born together with a sister. In those days, if twins were born, they had to marry each other when they grew up. This sister's name was Ntxawm, and she was the one who was supposed to marry the Emperor. But the Emperor saw a very pretty servant girl, and he married her instead. However, since she had still had no children after three years, he married another wife and waited three more years, but she did not have any children either. So he took a third wife, but still there were no children. Altogether he had waited for nine years. At last he married a fourth wife, and they all became pregnant together.

At that time Saub was not far away from people. And the Hmong were able to go up to heaven and come down again. So the Emperor said to his four wives, 'Tomorrow you four get up and cook rice to take with you and go up to heaven to ask Saub which of your sons will rule the land when I am one hundred and twenty years old.' So early the following morning they got up and cooked rice for the midday meal and went to see Saub. The first wife asked Saub about everything that had happened to them and which of the four wives' sons would rule and be Emperor in the father's place when he was a hundred and twenty years old. And Saub said, 'Whatever I tell you will surely be, and if I tell you, you will go mad.' But they assured him that they would not go mad. So then Saub said, 'All four of you will bear sons. But the sons of the first, second, and third wives will be just ordinary people like everyone else. The fourth wife will have three sons. The first will be the Emperor Kas Las Khej. The second will be the Emperor Tswv Ntxhw. The third, Maum Dev. It is these three who will be kings when their father is a hundred and twenty years old. But Huab Tais Tswv Ntxhw and Huab Tais Maum Dev will only be the assistants. It is Huab Tais Kas Las Khej who will be Emperor when their father is a hundred and twenty years old.[2] So now all of you return and tell your husband what I have said.' And they started out on the way back.

Half-way back, the first three wives said to each other, 'The Emperor is not going to love us if we tell him about this. So let's lie to the Emperor, and tell him Saub said that the first wife's son would be Emperor when he is a hundred and twenty years old and reign for two thousand years, and then the second wife's son would reign for two thousand years, and then the third wife's son would reign for two thousand years, altogether six thousand years. And let's lie to the Emperor that the fourth wife is to have three sons who will reign for six thousand years after that.'

And so they lied to the Emperor, and told him Saub had said that if he let the fourth wife and her sons live with them, the country would be overtaken by war.

They advised the King to make a boat and put some food in it, blindfold the fourth wife and put her in the boat, so that the boat would take her to the other side of the sea. 'Only let us stay together', they begged, 'and then there will be no war.' And the third wife said to the others that that night the Emperor was to sleep with the fourth wife; 'You two help me. I have three jars of silver and three jars of gold. Let us take them and offer them to Saub so that he will help us.' Now some people like to say that Saub is a liar, because he lied this once. And when the last wife arrived she realized that the three wives had lied to the Emperor. But the Emperor had not wanted to believe them, and had told them to come back again the following day.

That night the three wives carried the three jars of silver to Saub, and the next night they carried the three jars of gold to Saub. After Saub had taken the silver and the gold, he had to say what the three women wanted him to say. So they told the Emperor that he need not go to ask Saub himself, since they could call Saub to come there. And that night the Emperor called Saub to come. Saub was the son of Txiv Nraug Lwj Tub, who had sent him to rule the four corners of heaven. So then the King asked Saub: 'Saub! I married my first wife and after three years had no children, so I married the second and third altogether for nine years but they had no children. Then I married my fourth wife, and they all conceived together. So I sent my four wives to ask you which of their sons would be king after me when I am a hundred and twenty years old.' And Saub told him just what the three wives had told him. And then Saub went back again.

So the King was furious with his fourth wife. 'You bitch!' he said to her, 'I thought you were going to bear me a good child, but it seems you are bearing me a bad one. I am going to put you in a boat, and let the waters carry you to the other side of the sea.' He summoned people to come and help carry her blindfolded to the boat and placed provisions in the boat for her, but she begged him not to put her in the boat, and said that her sons should stay with him to help him when there was war. 'If you do not listen to me and believe that my three sons will help you, then one day you will regret it,' she said. 'Please let me live with you.' But the Emperor would not listen to her, saying, 'You dog! I'm not going to listen to you! Your children are dogs, and evil,' and he ordered them to put her into the boat and put the boat into the water and let her go right away.

But she said, 'Oh King, if you don't believe me, and are really going to cast me away, heaven will send war to your country and you will lose your land. When that time comes, you will need my three sons to come and help, and your other sons will come to ask for their help. If they are to come, you must give them these embroidery ends as proof to bear with them, and when I see them I will let my three sons come to help you. But if not, then I will not let them come to help.' And as they put her into the boat and cast her off, she added, 'If your three wives want their sons to come, then let their sons divorce their own wives first before they come.' And that is why when people want to go to America they always have to divorce their wives first, and always divorce the first wife first. And so the boat floated to New York.

At that time Txiv Nraug Lwj Tub's fourth son, named Luj Ncig Ntuj, whose every step encircled heaven nine times, travelled to where the three elder wives of the King were, and they gave birth to three sons on the same day, and at the same time the fourth wife, who was on the other side of the world, gave birth to three sons, all of whom would be *Huab Tais* in the future because Saub had said so already. So old Lwj [*sic*] Ncig Ntuj went over there and provided them with food to eat. And there the three sons grew up.

Later the Emperor wished to go back to marry his sister Ntxawm who lived in Mongolia. That was the one he hadn't married before. If he had married her, the country would not have had war and we Hmong would not have lost our country and wandered to other countries. At the New Year Ceremony—which started because the Emperor's grandsons had given birth to Nkauj Ntsuab and Nraug Nas,[3] who were twin brother and sister, so that they had to get married when they grew up, and had been born on the thirtieth day, so that all the Hmong came together to play the *qeeb* and celebrate their New Year feast on that day—the Giant (*nyav*) Emperor also came along to celebrate with the Hmong, and said he wanted to play catch with Ntxawm.[4] The real reason he had come was because he had wanted to marry her. He had come from the USA. When they play, if either person cannot catch the ball, they have to pay a coin each time they lose. Long ago people only played for one day, but Ntxawm and the Giant King played for three days, so now everyone plays for three days. The first day they played, the Giant Emperor lost a lot to Ntxawm, so she promised him that she would marry him if he was able to win. And the third day she lost a lot to him, and was unable to pay him back, so the King put her under his arm and carried her off to his own country. When this was reported to the Hmong King, he was extremely angry. And the *Huab Tais* summoned his three sons and told them to go and take her back. But their three mothers told the three sons they would be unable to fight the Giant Emperor. The only ones who could do so were the sons of the Emperor's fourth wife, whom he had chased away to the other side of the sea. So the mothers told their sons to go and look for their brothers and ask them to come and help. And they packed their belongings and left.

They went on to the sea and they saw a great gun so they shot it to the other side of the sea where their brothers Huab Tais Kas Las Khej and Huab Tais Tswv Ntxhw and Huab Tais Maum Dev were living. When they saw that they knew that their elder brothers were shooting, so they shot back, and the shooting became a bridge across the sea so that the elder brothers were able to cross over to them. They were very happy and had a great feast together. Then the older brothers told the younger brothers how the Giant King had captured their Aunt Ntxawm, and how their father, the King, had ordered them to go and fetch her back, and would not let them be kings one day if they were unable to do so. And the younger brothers, the fourth wife's sons, said, 'Don't be afraid. We will help you. You cannot go by yourselves. We will go by ourselves because we can go by air, by sea, and even if the earth were burning we could still go, but you cannot go in these ways.' And they said, 'We will go, and we will leave you three housefuls of rice to eat while you wait for us, and when those are finished there are still three fields of rice left. When you beat the digger, if it is very loud, then it will be time for us to return. You must wait for us here. We will go by air (by flying)....'

The remainder of this legend, and indeed its general structure, closely resembles a famous Laotian legend known as the tale of Sin Say, who is regarded as a Boddhisattva (Abhay and Nhinn 1965). The resemblances really are most striking, since the Sin Say legend depicts the sovereign of the land reigning together with his beautiful sister, Nang Soumountha, who is stolen away and ravished by a solitary Gnak (from which the Hmong term *nyav* probably derives) rather as Sita was stolen from Rama in the Indian epic. The sovereign then, like the Prince Gotama, who was later to become the Buddha himself, abdicates his throne and goes in search of his sister, the jewel of his life. On his way, he meets seven beautiful sisters, whom he espouses and retains his throne in order to bear a son who will be able to defeat the Gnak. The theme of the seven sisters of the Pleiades is a familiar one in much South-East Asian literature, and in some Burmese tales where they are depicted as birds. All the wives conceive together, as in the above tale; the first wife gives birth to a gold-tusked elephant, and the second to a snail and a boy born with a bow and arrow. They are later known, respectively, as Siharat, Sang Thong (the Golden Snail), and Sinsay, who conquers through his own merits.

Fearing calamities due to these unnatural births, the sovereign banishes them all with their mothers, but they are helped by Pha-Inn (Indra in Hindu mythology) in their exile, and just as in the above account, the six other brothers come to get their help in rescuing their aunt when they grow up. Although the combination of elements is different, there is no doubt that these stories are identical in origin, and, moreover, that they have been for some time.

In a USAID document dated 1963 (USAID–RDD, June 1963), where an ex-monk named Thongsar Boupha gives his version of the 'Story of the Meo People' (which he claims to have had from an old Hmong named Sai Vue at Nam Chong, who had heard it from his father) a hundred-year-old hermit takes *clay* to make the first man and woman, named Phoua Sank See and Ya Sank Sar, who have a powerful son named Sin Say before being chopped into twelve pieces to form the different kinds of people. The document is prefaced by a USAID official named Bill Taylor, who vouches for its accuracy, and declares his belief that the Hmong must have originated from West Mongolia because their name sounds like 'Mongolia'. It may have been from such suggestions that current beliefs originated. The clay, however, seems to be a Christian motif. Sin Say is clearly the Laotian hero Sinsay (pronounced Xeem Xaiv in Hmong), and Phoua Sank See and Ya Sank Sar must refer to the aboriginal Mon–Khmer ancestral spirits ritually propitiated by the Tai rulers of Chiangmai under the names Phu Se' and Ya Se' (Phu and Ya being generational terms for ancestors) and by the Tai rulers of Luang Prabang as Pu Nyoeu Nya Nyoeu (Archaimbault 1964; the 'Y' is pre-nasalized in Lao).

Most remarkably, this semi-Buddhist *Jataka*, or Buddha birth-tale, is also a messianic legend, in fact the major messianic legend of Laotian culture, and the Hmong refugee camps also abound in a variety of legends about a Hmong emperor named Xeev Xais, who is quite clearly the same character as Sinsay. It is thus very probable that what we are witnessing here is a major overlap and coalescence of different messianic traditions, in the context of the refugee situation and resistance to the Lao authorities, between the Hmong Tswb Tchoj, the Lao Sinsay, and probably also the Chuang movement among the Khmu' (Smalley 1965; Ferlus 1979).

From a more critical point of view, however, what is more remarkable is the extent to which this Lao legend has been transformed into a Hmong legend through incorporating entirely Hmong ethnographic details, such as the exchange of embroidery, the counsel sought by the wives from Saub, and the traditional courting game of catch to which the Ogre's capture of the young girl is attributed, while retaining much of its original structure in such a way as to function as a whole as an explanatory device for the dispersal of refugee families. There is no sense conveyed in the story, nor was there by its raconteurs, that it was not entirely a Hmong one. And from the viewpoint of real history, we may say that indeed it is a thoroughly Hmong legend.

Yet, the fact that the structure and plot of this legend should have been adopted so transparently from the messianic tradition of a neighbouring culture forces one to ask whether the same may not have happened with the Tswb Tchoj legend, which as we have seen provides the ultimate paradigm of Hmong ethnic identity and has continually inspired the messianic leaders who have emerged among the Hmong, in the context of Daoist rebellions against the Chinese state in which ancestors of the Hmong must have been involved? In this case, our conclusions that, in order to differentiate itself from a majority culture, a minority culture is forced more and more to appropriate the very terms and symbols of that dominant majority, would be amply justified. What is certain, however, is that such traditions, and the legends in which they are embedded, have now 'become Hmong' in every sense, and thus provide continuing evidence of the strength and adaptability of Hmong oral tradition, its capacity to incorporate new elements and to fashion out of them the ongoing parameters of Hmong identity.

A similar sort of pragmatic eclecticism, which we have already observed in the case of the Flood legends (Chapter 8), was evinced by many Hmong magicians who claimed, in the survey site, to have learned their practices from the Karen, Chinese, Thai, and Khmu' as well as from Hmong. Thus, in terms of the theory of deteriorating knoweldge outlined in Chapter 8, the ethnic sources of knowledge do not matter, provided that such knowledge can 'become Hmong'. In a rather similar way, the children of other ethnic groups may be incorporated into the Hmong kinship system through the mechanisms of adoption and intermarriage. While the variables may be changed, the parameters of the system remain constant, and it is through the constancy of these parameters that the Hmong define their own ethnicity.

The way in which the Hmong are currently refashioning their own history, utilizing elements of their cultural background to provide a framework for changing social and material circumstances, can be seen most clearly in such legends and the terms they incorporate. In a similar way to the incorporation of a Laotian culture hero into the genre of Hmong *dab neeb* to validate the Hmong dispersion, the attempt to trace the origins of the Hmong to a homeland in Mongolia, the teachings of the three original Hmong brothers which explain why the Hmong refer to themselves as *Peb Hmoob*, and the emergence of the *Koom Haum* movement itself (a phrase probably best translated as 'the organization'), all provide examples of the attempt to come to terms with the present in the light of past, remembered or re-remembered, experience. As Benjamin (1979) puts it:

To articulate the past historically does not mean to recognise it 'the way it really was' (Ranke). It means to seize hold of a memory as it flashes up at a moment of danger.

He goes on to say:

Historicism contents itself with establishing a causal connection between various moments in history. But no fact that is a cause is for that reason historical. It becomes historical post-humously, as it were, through events that may be separated from it by thousands of years. A historian who takes this as his point of departure stops telling the sequence of events like the beads of a rosary. Instead, he grasps the constellation which his own era has formed with a definite earlier one. Thus he establishes a conception of the present as the 'time of the now' which is shot through with chips of messianic time.

Benjamin is here using a specific, theoretical sense of the term 'historicism' which I would not use. I would rather have said that it was history as a discipline which sought to establish such causal relationships, and that facts become *historicist* posthumously, through their relationship with the present conjuncture. Evans-Pritchard (1961) made a similar point when he remarked that history was not a succession of events, but 'the links between them'. But this is a matter of terminology. The image Benjamin uses to convey the kind of history he is talking about is the French Revolution, which 'viewed itself as Rome incarnate' and 'evoked ancient Rome the way fashion evokes costumes of the past'. In this and other cases, men were consciously using the past to fashion a new present for themselves. And this is surely precisely the function of many of the *dab neeg* considered here: seizing hold of a memory as it flashes up at a moment of danger, rather than recognizing the past, necessarily, 'the way it really was'. In this sense, such legends can be said to constitute a real, lived history, to which questions of proof, truth, and falsity become irrelevant, and it is in this sense that history, of such a 'real' (inhabited) kind, in that it seeks to illuminate, clarify, and formulate the foundations of the present (like the exchange of embroidery above) can easily be admitted into synchronic ethnographic study. The kind of history contained in such oral traditions is the very opposite of a raconteur 'telling the sequence of events like the beads of a rosary', but is precisely 'grasping the constellation which his own era has formed with a definite earlier one'.

Thus, it is not ethnographically relevant whether the above legend was adopted from a Buddhist tradition, whether the San-Miao banished by the Emperor Shun were the ancestors of the present-day Hmong to whom the same term is applied, whether geomancy was truly originally a Hmong or Chinese system, or whether the opposition of metals to other materials truly reflects the beginnings of the Iron Age in China. What is important is the resonances which such images evoke in the present context, and their implications for the Hmong sense of their own ethnicity and identity, and it is these we have sought to examine in establishing a concept of 'real', rather than 'true', or 'false' history.

We have seen this 'reformation of the past' (which I gloss for 'real history') most clearly, perhaps, in the *dab neeg* we have just considered, which seeks to account for certain aspects of the refugee situation, such as the commercial value of embroidery and the refusal of receiving countries to accept polygynous families, in terms of images borrowed from a neighbouring Lao tradition and thrown back on to the canvas of a thoroughly Hmong past. We have also seen it as clearly in the ending of the tale of Tswb Tchoj (Chapter 7), with the birth of the Hmong Messiah in America, where the movement of the legend from China to America is, of course, a historical one. The remarkable feature of the tale of Tswb Tchoj, however, and the reason why it was examined at such length, is that it is much more than an ordinary *dab neeg*: it functions (as the story of the two brothers in Chapter 8 does for the sighting of *loojmem*) as a *charter for social action*, a *programme* which is symbolically activated by the 'bearers' of messianic traditions among the Hmong when seeking authority for their own actions and status. It is because, as we argued in Chapter 7, it so primarily expresses the cultural and ethnic opposition of the Hmong to others, that it may be employed to articulate actual or current conflicts between the Hmong and members of other groups, which consequently adopt a messianic form. Here too, history is being 'recreated'.[5] And it is this sense in which history is 'recreated' (whether the history be of a 'true' or a 'false' kind) which it is, as we have seen, one of the primary functions of an oral tradition to facilitate.

Finally, let us briefly consider the situation of the Hmong refugees overseas, which reflects back on the major themes illustrated in the preceding chapters, as well as providing further examples of the radical reconstitutions of the historical past which are enacted in the course of the refugee situation. During the early 1980s, Yangdao (1982) estimated there were 60,000 Hmong who remained in refugee camps within Thailand, between 50,000 and 60,000 in the United States, 6,000 and 8,000 in France and French Guyana, with smaller numbers of 500 in Australia, or 340 according to Lee (1986), 200 in Canada, 100 in Argentina, and a smaller number in China (c.f. Table 6). UNHCR departure figures do not break upland refugees down by ethnic groups, but in 1983, the Office of Refugee Resettlement estimated there were 61,000–64,000 Hmong settled in the United States. Despite the severity of 3,000 years of conflict with arguably the most powerful state in the world, hardships under the French in Indo-China, and persistent attempts to colonize, assimilate, and subdue the Hmong by the Thai, Lao, and Vietnamese, probably the severest challenge yet to the cultural identity and ethnicity of the Hmong is now posed by the dispersion of the Hmong to the 'four corners of heaven' and their settlement abroad as members of post-industrial nations. Yet, it seems that, at least for these first- and second-generation migrants, the strong ethnic boundary mechanisms formed by Hmong society through its historical experience as a minority in China and Indo-China, have proved adequate to reconstitute the society in its new status as an ethnic group on a global scale. International visits by groups of Hmong to other groups of Hmong are becoming increasingly common, and this, together with the entry of the oral tradition of the Hmong into McLuhan's post-literate society through the extensive employment of the telephone and cassette-recorder, marks a new stage in the internationalization of Hmong consciousness as an ethnic group. We may be able to link this to the emergence of the Hmong as an 'ethnic group' proper, which Ross (1979) distinguishes from the stage of a 'minority group' on the grounds that the former entails a stage of *conscious* mobilization and redefinition of identity, which only occurs under modern conditions of development (cf. A. Cohen 1969).

Cases of legal cultural defence are becoming more common in the United

TABLE 6
Distribution of the Hmong in the United States*

Region I		Region VI	
Connecticut	200	Oklahoma	600
Rhode Island	1,300	Texas	150
Region II		Region VII	
None		Kansas	1,000
		Iowa	1,000
Region III		Nebraska	300
Pennsylvania	1,200		
		Region VIII	
Region IV		Colorado	2,000
Alabama	200	Montana	600
Georgia	200	South Dakota	200
Kentucky	200	Utah	1,000
North Carolina	200		
Tennessee	600	Region IX	
		California	14,050
Region V		Hawaii	150
Illinois	1,500		
Indiana	300	Region X	
Michigan	500	Oregon	2,000
Minnesota	11,000	Washington	1,000
Ohio	600		
Wisconsin	4,000		
		TOTAL	46,050

Source: Lao Family Community.
*As of September 1981.

States, and among those reported for the Hmong are cases of 'marriage by capture', which would otherwise be treated as rape (cf. Goldstein 1986). The clash of different sexual moralities implied by such pleas reminds one irresistibly of that outlined with regard to Buddhism in Chapter 4, yet in most cases the protagonists are too young to have directly experienced such incidents themselves in Laos or in Thailand. In such cases, history is clearly being reconstituted, or re-enacted, and in such a way as to mediate and articulate the *agon*, the tension, between past and present. I will give further examples of this.

One of the most interesting phenomena which have taken place among the Hmong refugees since 1980, which has also affected the Hmong in Thai villages and refugee camps, is the discussions which have taken place over how to pay the traditional bridewealth expected at weddings, now that no silver and little else is available which might act as acceptable bridewealth. It has even been seriously discussed as to whether a written contract registering the exchange of a bride between two descent groups might not be an acceptable substitute for bridewealth payments. Although such suggestions have proved difficult to implement owing to the localization of particular descent groups and the wide dispersal of the clans of which they are members, they do mark a very sophisticated stage in the understanding of the value of bridewealth as the record of a transaction rather than a payment for it, and further provide an example of the way the Hmong have become, though the refugee experience, active participants in the reconstitution of their own society. This was also evident in a Hmong Studies Conference, held at the University of Minnesota in 1983, which was attended by a great number of Hmong refugees, and in the growing volume of Hmong publications, of which Bliatou's (1982) study of the nocturnal sleeping death syndrome among Hmong refugees is only one example (cf. Lemoine 1982b).

The persistence of ethnic boundary mechanisms may be seen in the large-scale secondary migrations which took place among Hmong refugees in the United States following their original settlement in widely dispersed areas. Thao (1982) distinguishes between secondary migration for regroupment and secondary migration for social betterment, and gives the example of hundreds of Hmong who moved to Portland after an old man was murdered. While certainly such migrations are of a very novel type in the United States nonetheless they retain certain features of traditional Hmong solutions to misfortune and scarcity in their life as shifting cultivators, as we saw in the case of Suav Yeeb's household described in Chapter 5, and also in the village in the survey site which relocated after a series of disasters.

Medical and linguistic problems remain as pressing for the Hmong refugees overseas as we have seen them to be in the case of the village Hmong in Thailand. The difficulties of communication in the language of the dominant majority remain insurmountable for many first-generation refugees, and the high rates of Hmong unemployment reported in certain areas are often linked to inadequacies of language training programmes. This, in turn, has enhanced the sense of Hmong ethnicity as a 'bounded entity under threat' (Irvine 1982), which is transmitted to second- and third-generation settlers, and also contributes to problems in dealing with different systems of medicine. Given the importance accorded to health in an ordinary White Hmong village, described in Chapter 5, it is perhaps not surprising that many Hmong refugees should have taken abroad with them medicinal green herbs which formerly grew in patches about the outskirts of the village or had to be searched for in the recesses of the forest, and that they should have planted these green herbs in and around their new homes in France and the United States, and searched out herbs in these new countries which were similar to those grown at home. Several cases have been reported of fatal cases of misunderstanding between Hmong refugees and Western-trained doctors (Evans 1983, p. 110; O'Neill and Rugoff 1982) which result from the conflict of different conceptual systems associated with different languages.

Certainly the trauma of leaving Laos, the long exile in the 'no-man's' land of the refugee camps, the often incomprehensible flight to a new home, and the difficulties of adjustment to radically new social systems, have all contributed to mental and psychological problems among the Hmong refugees which are often closely related to social problems and may partly explain the suicides and nocturnal deaths among them (cf. Lemoine 1982b).

At the same time, traditional methods of dealing with these and other health-related problems are also under threat, not only from different medical practices, but also from Christianity, which continues to win adherents from the Hmong in the refugee camps, where educational and relief projects are often controlled by Christian agencies, and in their new homes overseas where local church organizations may sponsor arriving refugees. Shamanism continues, for example, both in France and the United States, but often in a drastically modified form owing to restrictions on the domestic slaughter of animals or the neighbours' complaints. It is surprising, however, to what extent the abandonment of traditional medical practices has not entailed the loss of a fundamental faith in them.[6] For example, the absence of shamanism or domestic worship in American households is sometimes attributed to the structure of the apartments, which lack the house-posts which are the abode of household spirits. And when I asked why the Hmong had stopped wearing their traditional silver collars, or *xauv*, believed to bind the wandering *plig* soul-substance more closely to its home in the body, expecting to hear that they had been found to be inefficacious, I was told that in a less densely forested country such as the United States, there were fewer wild forest spirits which might capture and endanger the wandering *plig*. Clearly these are secondary elaborations of the type described by Evans-Pritchard (1937) for the Azande. Yet such explanations remarkably reflect the strength and vitality of a conceptual system to absorb and accommodate change, which it has been the argument of this study is facilitated by the adaptability and flexibility of a living oral tradition.

Apart from the increasing divisions between Christian and non-Christian Hmong among the overseas refugee population, traditional clan organization has, if anything, been reinforced through the experience of migration abroad, as the discussions concerning bridewealth show. It may be that Hmong clan structure will follow the pattern described for Cantonese lineages by Watson (1975), as in some areas clans have formed lending pools for medical insurance and the purchase of luxury items, such as television sets and cars. At the same time, however, there is a growing aspiration towards a new form of social organization which we may call of a non-segmentary nature (cf. Ajchenbaum and Hassoun 1980). This is observable in the opinions often voiced by refugees that the dispersal and lack of unity of the Hmong, and their defeats in Laos, are directly due to the problems of feuding and disputes between different descent groups and different clans. The legendary figure of the Hmong Emperor represents a mythical attempt to overcome this inevitable opposition between clans, and the *koom haum* movement, with its prophecies of the unity of the clans, a more actual attempt to transcend fission and attain fusion. The Lao Family Organization, a nation-wide organization set up in the United States to represent Hmong interests, may represent another attempt of this kind. As we have seen in the course of this work, divisions between descent groups and clans are most clearly represented in the course of mortuary rituals dedicated to divergent lines of ancestral spirits. These distinctions are physically represented in the placing of burial sites according to the geomantic principles of the system of *loojmem*, so that not only the cleavage of the Hmong and Chinese, but also the feud between Lis and Lauj clans which led to the political polarization of the Hmong in Laos, and hence to the situation of the refugees, is traced to the adjacent burials of relatives. However, such adjacent burials are difficult to avoid in church graveyards overseas, unless cremation is practised, which leaves nothing in the grave to be reincarnated. Reincarnation can only take place once everything buried in the grave has perished, as we saw in Chapter 6.

This situation has caused much fear and anxiety among elderly Hmong refugees, who prophecy future misfortune from their inability to act in accordance with geomantic principles in their new homes overseas, or even to perform the proper burial rites for relatives, since it is difficult to keep bodies in the house while the necessary funeral rituals are performed. Particular distress has been caused by the autopsies performed on victims of the nocturnal death syndrome, since damage to parts of the body is believed to be replicated in the next life: in some cases, relatives have had to remove metal staples from these bodies since, as we have seen, imperishable metals are particularly inauspicious. These stories have reached the refugee camps in Thailand and contributed to a no-show rate among refugees accepted for resettlement of 40 per cent at the time I visited the camps. However, further proof of the way the Hmong are examining, reflecting upon, and criticizing their own social structure is given in the suggestions which have been made to overcome these problems, that Hmong temples should be set up where proper rituals could be performed for the deceased. Ultimately, this would entail the obliteration of clan distinctions which forbid the performance of funerals in homes other than those of the descent group (and thus clan) of the deceased. Here we can sense the relevance of Carr's (1961, p. 129) comment that 'History begins when men begin to think of the passage of time not in terms of natural processes . . . but of a series of specific events in which they are consciously involved, which they can consciously influence.'

The dread and anguish resulting from the encounter with and forced adoption of the very different mortuary customs of their host countries should not be dismissed lightly, since they arise from probably the most coherent metaphorical system ever invented for the expression of man's relationship to the cosmos. Disorders registered in terms of this geomantic system relate to far more fundamental disorganizations of that relationship. Although many Hmong overseas now recognize the need to conform to the established norms of the societies in which they have resettled, still it is hard even for these to reconcile these norms with their own up-bringing and values—and there is, I think, an important argument for the recognition and respect of values which may be very different from our own. The cultural atrocities represented by Western mortuary practice, as a whole, and by the autopsies, in particular, are the final indignity, the last outrage, on a par with the Chinese grinding Tswb Tchoj's bones into powder. And whether a new King, that is, a new cultural resurgence of power and vitality, will arise, will remain to be seen.

1. Some 46,000 Hmong remained in five refugee camps at the time of my visit, out of a total population of Laotian 'hilltribe' refugees in those camps of 48,500 (UNHCR, Bangkok).

2. These names refer, respectively, to the crocodile, the elephant, and the bitch.

3. The first couple.

4. The courting game played between unmarried boys and girls at the New Year. *Ngac*: originally *Nago*, like Thai *naag* (or Burmese *nat*?), great subterranean serpents prior to the advent of Buddhism.

5. Cf. Hobsbawm and Ranger (1983). The distinction between 'custom' and 'tradition' drawn by Hobsbawm in this book is a thin one, however, for there is surely a sense in which 'tradition' has always been reinvented, and a more precise sense, which has both scientific and artistic correlatives, in which it cannot always be decided what the difference between invention and discovery is, as in the 'discovery' of the principle of gravity. Cf. Josipovici (1971, p. 194) on the same tension between invention and discovery in Renaissance art.

6. This was also the case with the explanation given for Txoov's death in Chapter 6.

Epilogue

THIS book has dealt throughout with the relationship between the worlds of the possible and the realm of the actual. Thus, the actual village situation was defined, in Part II, in terms of polar, alternative oppositions; between a mixed economy and a permanent-field economy of non-poppy crops, between loyalty to the Thai state and support for the CPT, between the adoption of Thai Buddhism or foreign Christianity. Obviously, these oppositions were not all of the same order, yet they all had one thing in common, which is that none of them described the village adequately. For the village itself was neither completely traditional in the economic, political, or ritual senses, nor was it Christian, Buddhist, firmly politically aligned, nor yet a permanent-field cash crop economy. Yet all these extremities were present in the village, in a latent rather than a manifest form, and it was out of the interactions between their polar oppositions, in the choices and decisions which confronted the villagers in their economic, socio-political, and ideological lives, that the *actual* situation of the village was defined.

The village thus provided a focal 'field' from which the various conflicts and alternatives, found there in latent or microcosmic form, could be examined within a wider historical and geographical context. While the village provided this kind of arena for the contest of wills, choices, and decisions (described as the 'kernel' of this study in Chapter 5) which defined everyday existence for the White Hmong, these choices and conflicts themselves had their roots in historical conditions, and their resolution was constantly referred to oral traditions of the past which expressed a particular view of history. It was this which defined the ethnicity of the Hmong, as it was experienced by them, and, through affecting the current process of decision-making and choice, influenced the course of *future* action.

For this reason, when we turned, in Part III, to a consideration of the past through villagers' eyes, we were throughout dealing with a realm of the potential rather than the 'has been', the 'might have been' rather than the 'was'. Rather than going into the historical dimension overmuch, I sought to bring history into the field of inquiry, by basing it on the historical *consciousness* of the actors, with reference to the *choices* which confronted them. Thus, whether the Hmong *truly* once had Emperors of China, controlled the lands of China, had a form of writing for their own language, and a state of their own, did not concern us. Those questions were bracketed. What *did* concern us, throughout, was the *conditional* nature of utterances to the effect that once they had done so; that is, 'if it *had not been* for X, then we *would have had* Y', which becomes, in that history which is real for being remembered, 'we *had* Y, but *because* of X we lost Y'.

Steiner's study of the problems of translation has examined the way in which such conditional statements, and the general capacity of language for misinformation and 'counter-factuals', express that 'alternity' to a given situation which animates the forward motion of history (Steiner 1975). For, in an important sense, the worlds of aspiration and possibility expressed in these views of the past had neither been exhausted by that past, nor were they by the present, which was characterized by precisely the same sort of economic deprivation and ethnic discrimination which the mythological statements about the past sought to explain. That is, the objectives expressed by such aspirations remained in the realm of the *potential*, not in that of the actual. And it is this potential which draws us towards a future, which is itself utopian.

So that when we turned to the future, or at least to that future expressed in the messianic longings for a better society most clearly manifested at present by the Hmong refugees, we found that precisely similar concerns were expressed as those which had been projected into the past, arising out of current medical, educational, occupational, and territorial deprivations. Utopian aspirations of this kind fall into the category of the *Noch-Nicht* or 'Not Yet' which Ernst Bloch tried unsuccessfully to introduce into mainstream Marxian thinking (Hudson 1982), just as the category of the 'might have been' in the past, represents a future which, although not fulfilled in that past, still remains a potentiality of it.

This category of the *Noch-Nicht*, freely translated into the past as the category of the 'might have been', has allowed us to provide a negative definition, a via negativa, of the actual situation of Hmong society in the 'real' (but yet unfinished) world, through a concentration on what it was lacking, or its deprivations (in the past as in the present). The fact that the *dab neeg,* or legends, which have formed one of the primary materials for this study also seem to define White Hmong ethnicity in terms of what it is not, may lead us towards a new vision of the function of ideology; not necessarily obscurantist (cf. M. Bloch 1977) but, in its inversion of the real order of things, depicting a better, and a more humane, society.

The most general conclusion to emerge from the book is that the Hmong have, in the course of their history, made use of certain oppositions to characterize their differences from other ethnic groups, and consequently define their own identity, but that in the course of so doing, these same oppositions (for example, between swidden and permanent forms of agriculture, or having an Emperor and not having an Emperor) have become internalized *within* the categories of Hmong society, to such a point that Hmong society itself contains all the oppositions which were used to characterize its differences from other (neighbouring) societies, so that identity has become difference, and difference identity. In other words, the only terms available to the Hmong to distinguish themselves from other societies, were precisely the terms and symbols used by those societies themselves, so that symbols which were originally used of difference, have become symbols of (Hmong) identity. In effect, this is a process of negative dialectic: the more we say we are *not* like them, the more we have to define ourselves with reference to, and by, them. If one can assume the *primacy* of difference, then it seems evident that the terms in which the 'Hmong' were forced to differentiate themselves from the 'Chinese' had from the start to be 'Chinese' ones, which may explain the crucial importance of the Tswb Tchoj legend and the system of *loojmem* in articulating such differences from the Chinese, and their continuing importance in articulating relations with other state traditions, as well as the *negative* formulations of Hmong identity (*absence* of states, kings, writing, power, and so on).

To the extent to which it is *not* possible to assume the primacy of difference, however, between the two ethnic groups represented by these ethnic categories, it became necessary to distinguish a 'real' history, which implies a *conscious*, choice-full selection from the sum total of past events and possibilities, from a 'true/false' history, in which particular events can be proven or disproven to have taken place at particular loci of past time, by reference to monuments, *written* sources, or other kinds of evidence considered more authoritative than mere speech. As Carr has put it, 'It is only through the ever-widening horizons of the future that an approach may be made to "ultimate objectivity" in historical studies' (Carr 1961, p. 117).[1] And it was in oral traditions and individual recollections of the past and their interactions with other kinds of evidence, that we revealed a real, that is, a *felt*, history, which concerned the past as much as it did the future.

The fact that those states from which the Hmong have sought to differentiate themselves have been invariably represented by more powerful and dominant societies with which the Hmong were in a relationship of subservience and subordination, raises further important questions about the *options* available to a minority culture to distinguish and segregate itself from, or to legitimate an already existing separation and segregation from, the members of such a more powerful, majority or 'host' population. Obviously, it was not possible in this work to examine this issue from the comparative basis it deserves, but through including material collected in the refugee camps of Thailand where many Hmong now languish, it is hoped to have thrown some light on this problem, since it is in such artificial environments that the conflicts between minority and majority populations, which we have examined at the village level as well as in formulations of the historical past, become, if not most acute, at least most clearly apparent.

1. On the selective nature of history, and the ambiguity of the term in referring both to a process and reflection upon it, see also Croce (1941), Collingwood (1946), and Berlin (1954).

Appendices

APPENDIX 1
Previous Work on the Hmong

SINCE the immediate sources on the Hmong are so numerous, I shall not deal here with general Northern Thai ethnography, nor with the literature on other groups of shifting cultivators in the same ecological region (including Northern Thailand, eastern Burma, southern China, and the northern parts of Laos and Vietnam). The first general ethnographic surveys of the Northern Thai region were those undertaken by Hanks and Sharp (1964), and Lebar, Hickey, and Musgrave (1964), while general compendia to date comprise Young (1961), Kunstadter (1967), Hinton (1969), Keen (1978), Walker (1981), and McKinnon and Bhrukrasri (1983). However, I should like my own work to be considered within the general context of the debate on ethnicity which has characterized much of the regional ethnography since Leach (1954, 1960–1), followed by Lehman on the Chin (1963) and Karen or Kayah (1967a, b, 1979), the debate between Moerman (1965, 1968a) on the Tai-Lue from the Sipsong Panna region of China and Naroll (1964, 1968), and Hill (1982) and Forbes (1987) on the Yunnanese Chinese, who have acted as intermediaries in the region between upland-dwelling groups and lowland populations (cf. Leach 1960–1). Here I shall concern myself solely with substantial modern works on Hmong society in European languages. The only major omissions from such a consideration will be Ruey Yih-fu's and Lin Yueh-hua's work on Miao groups in China. I am thus excepting articles arising either out of such substantial work or less substantial work, 'refugee literature', and older or unreliable materials (but see Bibliography).

Early Sources

The earliest substantial sources on the Hmong are those of Savina (1930) and Bernatzik (1947, trans. 1970). Savina's *L'Histoire des Miaos* is the work of a Catholic missionary who worked for many years with the Hmong of Tongkin. A large part of the book contains a vocabulary in an early transliteration, while much of the remainder is devoted to a basic description of the peoples and their conjectured history in which Savina tries to find echoes of Biblical themes (for a critique see Graham 1937). Although it cannot be treated as a serious source of ethnographic information on the Hmong, it does contain information which is useful for comparison. The account of Paj Cai's revolt, witnessed by Savina, is of particular interest. Bernatzik's *Akha und Meau* contains the results of a field expedition undertaken among the tribes of 'Further India' in 1936–7, and attempts a comparative study of the Hmong and Akha of North Thailand. Although that part of the book which deals with the Hmong contains a large number of inaccuracies, it does contain some basic information on material culture at that time. Ethnographically speaking, the work cannot be treated as reliable. For an impartial indictment of his field techniques, anthropological outlook, and general conclusions, see Skinner (1956).

Green Hmong Materials

Serious modern ethnography has been largely devoted to the Green Hmong, and where it does concern the White Hmong, with one exception (Moréchand, see below), deals only with socio-economic analyses and makes no attempt to tackle cultural issues. Thus, the only two published ethnographies of the Hmong are those of Lemoine (1972a) and Geddes (1976), both ethnographies of the Green Hmong (see Bibliography). While Geddes' material is interesting since it does concern Hmong of the North Thailand area, and he had a lengthy if intermittent acquaintance with the area (employed by the UN to conduct their first survey of the opium-producing areas in Thailand and later to establish the Tribal Research Centre, the government's main research wing in the area), only half of the book attempts an (idealized) ethnography, while the entire second half is devoted to a study of the economic effects of opium production. His work on the Hmong, unlike his earlier work on the Dayak, was thus inevitably overshadowed by his official concerns, while the ethnographic parts of it suffer from imprecision and overgeneralization and do not argue a very deep acquaintance with Hmong society and language as actually lived (see some of Lee's criticisms, 1981). By contrast, Lemoine's ethnography, also of the Green Hmong (termed 'Blue' by Geddes), to which he brings a thorough knowledge of the Chinese sources and an equal familiarity with Yao and Lao cultures, is excellent. While the ethnography itself is concerned almost entirely with material culture and social organization (an account of Green Hmong ritual and belief is planned for a later volume), several articles, in particular three on their funeral chants (Lemoine 1972b), do contain valuable information on Green Hmong culture, and I have referred to these where appropriate in the book. I should add that it was Lemoine's preface to Feuchtwang's study of Chinese geomancy (1974) which first alerted me to its practice among the Hmong of Laos (where Lemoine worked), and inspired me to analyse its application among the White Hmong of Northern Thailand.

   Geddes worked in collaboration with a Thai assistant from the Tribal Research Centre, Nusit Chindasri, and in a preface to his own work explains that they had divided the work between them, with Chindasri concentrating on the cultural and religious aspects of Green Hmong society and Geddes on the economic aspects. Chindasri's work has appeared in paperback form (1976) and, while it does make a serious attempt to provide a detailed description of the major aspects of Green Hmong religious life and belief, unfortunately there is little or no attempt at analysis, as his introduction notes. He admits that despite the division of labour between himself and Geddes, 'our main concern, however, was the economic pattern'.

White Hmong Materials

There are four further major works to consider, three of these specifically on the White Hmong. The first of these is an (unpublished) doctoral thesis prepared for the Advanced Research Projects Agency of the US Department of Defence, by Binney (1971), whose fieldwork was part supervised by Geddes. This provides a detailed, careful, and well-researched account of White Hmong *economic* organization in North Thailand, with some account of the basics of White Hmong social organization. It also shows signs of some familiarity with the language (although misspellings are rampant). However, it makes no attempt to deal with any of the cultural or historical issues which are the concern of my own book.

The only ethnographical work which does concern such cultural and religious matters among the White Hmong is contained in two lengthy articles in *BEFEO* by Moréchand (1955, 1968), who worked in Laos and northern Vietnam. While White Hmong is a surprisingly homogeneous society (see Bertrais 1978), there is no guarantee that what is true for the White Hmong of North Vietnam is true for the White Hmong of Thailand. Moreover, Moréchand's two articles are strictly confined to a study of *shamanism*. While astute and highly professional in description and analysis, even in the circumscribed sphere of shamanistic rituals, some of the practices and beliefs he describes are simply unknown among the White Hmong of Thailand.

The only other anthropological work dealing specifically with the White Hmong, besides Moréchand's two articles and Binney's economic analysis, is an (unpublished) doctoral thesis actually by a White Hmong *emigré*, who also studied under Geddes at Sydney University. This is Gar Yia Lee's 'The Effects of Development Measures on the Socio-Economy of the White Hmong' (1981). Again, this is a highly specific work, concentrating on socio-economic issues. Moreover, it does not deal with the overall impact of agricultural, religious, social, and educational government programmes (as my own MA dissertation, 1979, attempted to do) but is confined to a study of the impact of the UNPDAC crop-replacement project on one White Hmong village. In an opening chapter, a broad overview is given of 'Social Structure and World View' which contains some account of White Hmong beliefs and custom, but this is not undertaken in any depth, and even contains some serious inaccuracies (for example, the identification of Saub with Ntxwg Nyug, p. 76, two completely different cultural figures, one a benevolent *deus otiosis*, the other a malevolent or, at most, indifferent Lord of the Otherworld). Lee does not attempt, however, seriously to deal with these sorts of cultural issues, and the thesis remains primarily an economic study. Thus, there are three specific works on the White Hmong: Lee's study of the effects of the UNPDAC crop-replacement programmes, Binney's study of traditional shifting cultivation, and Moréchand's two articles on the White Hmong shamanism of Laos and North Vietnam, all useful sources which I have referred to wherever necessary or appropriate, but which make no attempt to deal with the subject matter of this book.

One final study of Hmong economic organization remains, which is that of Cooper (1984). Cooper studied mixed Green and White communities distant from each other on the grounds that it was impossible to isolate a typical community. For my own purposes, I found it preferable to confine study to a single period in a homogeneous location, and follow the networks of interaction outwards. One of his conclusions is of direct interest to the concerns of this book, which is that, on account of the lack of incidence of Hmong 'becoming Thai' (p. 311), 'the concept of Thai-ization through "culture contact" has little application to a study of change among the people of Northern Thailand'. While I would agree that what Thai-ization has occurred among the Hmong has been minimal, I have found it more interesting to inquire into the reasons why this has been so.

Cooper, who coincidentally selected not only one of the villages previously studied by Geddes but also the same 'general study areas ... of settlement and economy', has very much, like Binney and Lee, produced an analysis of the system of shifting cultivation and (like Lee) the effects of government intervention upon it—in effect, an analysis of the emergent class formations brought about through the adoption of permanent-field agriculture. However, as we have seen in this work, not only has this occurred on a limited scale in the past decade but (as Cooper himself notes) there is strong evidence that permanent-field agriculture has always played a limited part in the historical past of Hmong agrarian practices. In my own approach, I have preferred to see the conceptual opposition between state and stateless people as continuing to define the religious, political, and economic alternatives for the White Hmong of Thailand. However, in general, I have not attempted either a socio-economic study of the type provided by Binney, Cooper, and Lee, nor an analysis of shamanistic ritual of the type provided by Moréchand. I thus hope that this work, which was from the start intended to be a 'micro-study' of geomancy, messianism, and the perception of time (the latter transmuted into a concern with the status of historical consciousness) will, far from treading on any toes, make up for a serious gap in the current ethnography of the White Hmong and, in conjunction with those works already existing on the Green Hmong, contribute to our general understanding of the Hmong.

Missionary and Other Materials

In conclusion, I should mention the extremely valuable collections of Hmong ritual materials which have been prepared in Hmong and French by two missionaries who worked with the Hmong in Laos and Thailand, respectively. The first is a mighty and as yet unpublished compendium of White Hmong wedding songs and procedures by Father Bertrais (also responsible for one of the only two White Hmong dictionaries in existence!). Although no ethnographic analysis of these is provided, they constitute a very useful source of information on different kinds of marriage and courtship. Four further works have been published by Father Mottin: a brief history of the Hmong (in English); a collection of folktales (*dab neeg*) of the same generic type as those which I collected in the village (and which I have cited in comparison wherever appropriate); a major collection of shamanic chants; and a compendium of New Year celebratory rituals and customs, which include description and comment. These sort of materials are invaluable in comparison for any ethnographer, and to these should be added Graham's 1954 collection of Ch'uan Miao songs and legends from China which provides interesting if not thoroughly reliable material (Graham almost certainly worked in Chinese). Finally, on 'Miao' groups of China, Inez De Beauclair (1960) should be consulted, while Yangdao (1975) has provided an illuminating study of socio-economic change among the Hmong of Laos from a geographical perspective. Other useful background materials include Lombard-Salmon's major historical study of the Miao in Guizhou province of China during the eighteenth century (1972), and Eberhardt's 1982 work on Chinese minorities.

Thus, it can be seen that materials on the Hmong are abundant, although the present work does not overlap with the concerns of previous researchers in any way. Through the medium of ethnography, I have tried to grapple with certain problems in the cultural aspects of White Hmong society, and to relate them to my own findings on the basic economic, political, and ideological structure of White Hmong village life as I found it in the survey site. I hope that work on the White Hmong will continue, for there is still much more to be done.

APPENDIX 2
Household Composition of Nomya

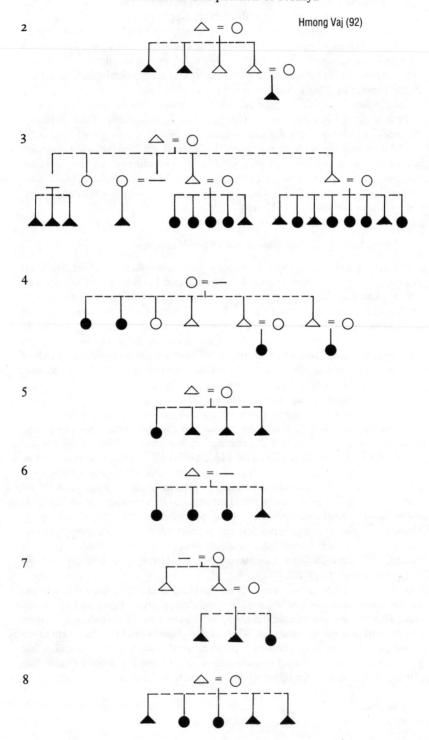

Hmong Vaj (92)

9

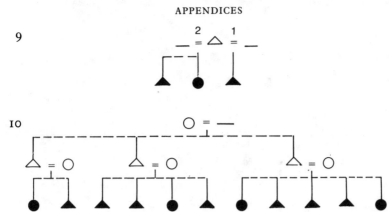

10

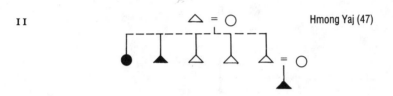

11

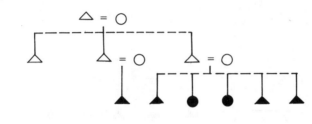

Hmong Yaj (47)

12

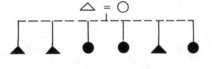

13

14

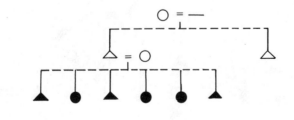

15

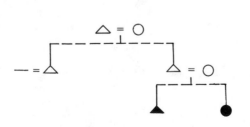

16 Hmong Xyooj (68)

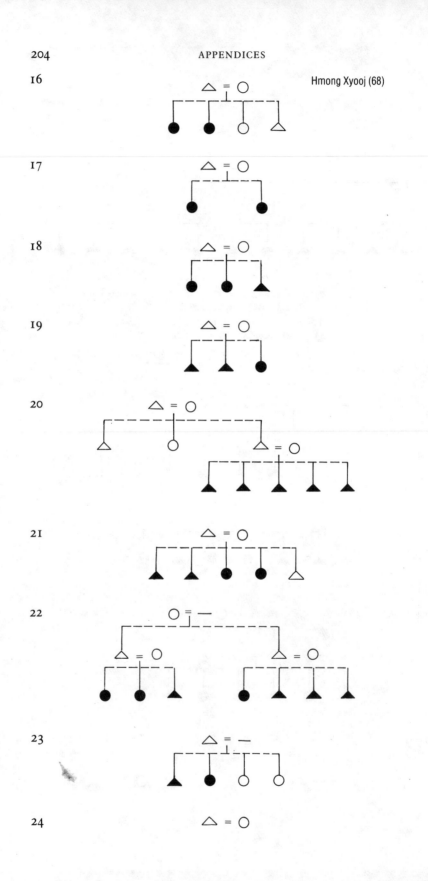

17

18

19

20

21

22

23

24

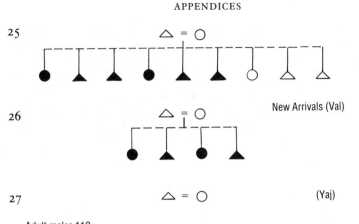

25

26

New Arrivals (Val)

27   (Yaj)

Adult males 112
Adult females 99
▲   63
●   52

Notes

I have indicated children of fourteen or under in black, and for the sake of clarity have occasionally indicated a deceased or deserted member of the household by a dash. It will be seen that two-generational households are the norm (eleven three-generational ones and one single-generational one, the latter a childless old couple), and that out of the three-generational ones, five involve widowed mothers living with their sons, and the descending generations of all of them are composed of young children. The great difference between the labour capacities between households is also apparent, with a household of two at one end of the scale, and of twenty-five at the other. It will be seen that while Vaj outnumber Yaj and Xyooj separately, they do not outnumber them both together. This partly explains the insecurity of the Vaj headman's position, and why the old Yaj man of the village still retained some power, since his was the casting vote in any conflict with the Xyooj. The sample is too small, however, to draw general conclusions from, and many other factors need to be taken into account if the household composition of the village is to be properly understood. It would be misleading to make a straight comparison on the basis of the data given above (for example, to deduce that nuclear households rather than extended ones were more prevalent among the White Hmong, or that the labour differences represented permanent wealth stratifications) since all the forms represented typify stages in a domestic cycle in which eldest sons left parental households with their children to found new households, and daughters married out. To understand this, the main relationships between the different households as well as biographical data need to be taken into account.

For example, the composition of household number 3 is very much explained by the series of misfortunes and mésalliances described in Chapter 5. A daughter had just married out of household number 2, leaving only brothers. Household number 17 had remarried in middle age, and because his two wives did not get on maintained another household in a different village. Descent relations (see kinship chart in Fig. 4) included households numbers 7 and 8, whose eldest male members were brothers who had previously lived together, and similarly households numbers 9 and 10, in the Vaj group. Household number 6 was headed by a man whose father had been the younger brother of the father of household number 5, and after his father's death he had been brought up by his uncle in household number 5. Households numbers 11, 13, and 15 in the Yaj group all represent splinter groups of household number 12, since household number 15 was headed by number 12's younger brother, while the heads of households numbers 11 and 13 were his sons. In the Xyooj group, household number 19 was headed by the eldest son of household number 16, while numbers 21, 22, and 24 were all headed by brothers. And household number 26, which moved into the village while I was living there, was the younger brother of Vaj household number 2 (see Chapter 3). So that a constant process of fission and aggregation is taking place, which a cross-section does not represent very adequately. In terms of affinal links, the most important links to bear in mind were that daughters of Yaj household number 12 had married into Vaj households numbers 2 and 10, and Xyooj household number 18, and the same household's eldest son (11) had married the sister of Xyooj households numbers 21, 22, and 24. Indeed, in kinship terms the village reads as a Yaj core cluster to which the Vaj and Xyooj families had attached themselves separately and through different links. Thus the old man of the village and the ritual head of the Vaj descent group (3) were (half) brothers-in-law. The very slightly higher proportion of males to females remains relatively constant in the under fifteen age-group.

APPENDIX 3
Population Structure of the Focal Village

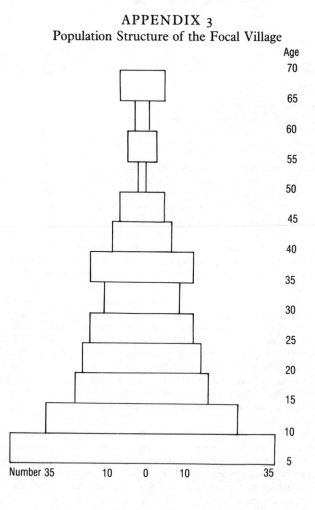

Notes

It was possible to gather highly accurate age statistics although ages are not normally reckoned accurately. In this I was greatly helped by having an assistant who had been educated outside the village, but had grown up within it, and consequently could cross-check villagers' relative assertions with his own knowledge of dates and major events outside the village as well as within it.

1. *Mortality*

I was not able to gather accurate data on mortality, since villagers were extremely reluctant to discuss family deaths, and it is, in any case, prohibited to discuss death in the house, during daylight hours, etc. However, mortality figures did emerge from the very extensive genealogies I collected, which I cannot include here as they would require some fifty pages. What is very certain is that the figures are an underestimate; they could not possibly be overestimated given the sort of feelings I have mentioned, while it is probable that many were not given out of natural reluctance, a preference to forget, and fear of supernatural sanction. No adult deaths occurred in the focal village while I was resident there, and the figures I have generally relate to child deaths. There were five infant deaths during my stay in the village, at least one still-birth and several miscarriages. Unfortunately, for the same reasons, it was impossible to distinguish, in the genealogies, still-births from infant deaths, although I was aware of the value of doing so. A very typical emic picture of the pattern of childbirth was given me by the wife of the head of Yaj household number 12, a herbal expert, who told me that she was a fortunate woman, having had five sons, five sons-in-law, and five dead. Daughters only counted in as much as they had gained husbands. The mother of household number 2, besides bearing five fine children, four boys and a girl, had given birth to an eldest son who had died in his first year, and having had three miscarriages, could have no more children. Household number 3, as indicated in Chapter 5, had five surviving children (including two out-married daughters) out of an original ten, which included three teenage deaths, one child who had died aged three days, and another who had died 'very young'. This is the sort of data which were gathered unsystematically, but it implies that a third to half of the children born do not attain maturity.

Since it is not comprehensive and probably underestimated, I will not tabulate the following information, but will briefly rehearse it here. The death of the wife of household number 6 had taken her child with her. Of the three sons born to household number 13, one had died aged about five months, while four daughters lived. Household number 21 had suffered two miscarriages out of five sons (since unborn children are assumed to be male), and one of their three daughters had died aged four. Of the three sons born to household number 19, one had died in his first year, and two of their three daughters had also died, one immediately after birth. Two of household number 8's five sons had died in infancy, leaving two daughters. Of household number 20's three sons, one had been killed in an accident during his teens, and of seven daughters two had died of an infection, aged about seven. One of household number 23's two sons had died aged only a few months, leaving him with three daughters. In household number 22, two out of five sons had died aged a year and less than a year, and of four daughters three had died sick aged a year and less than a year. Of household number 17's three daughters, one had died at about a month old. The head of household number 16 said three of his six sons had died, two a year or more old, one immediately after birth, while one of his six daughters had died of a stomach complaint when she had been about four. His wife contradicted him, however, saying she had had nine daughters altogether, two more had died when they had been a couple of years old, and one had died after birth. In general, it was much easier to talk with the women about this.

2. Second Marriages

Since the Hmong are usually described as having a polygynous family system, it is important to note the rarity of actual polygyny among the White Hmong. The second son of household number 3 had remarried *after* his first wife had left him. The house also contained his younger brother's young widow, who was expected to remarry shortly, with her infant son (she could only have remarried a *younger* brother of her husband). The head of household number 5 had married a widow. The wife of the head of household number 6 had died of dysentery a year before I came to the village, and with four children her widower was expected to remarry. The younger son of household number 7 was being forced to take a second wife (the daughter of household number 16) by the girl, who was pregnant, and her parents, although he was only twenty-one years old and his first wife had threatened to commit suicide if he did so, supported in her opposition to the second marriage by his mother who lived with them. This dispute continued through the period of fieldwork but, although there was considerable male sympathy for him, according to customary law he would have no choice but to take the second wife if the pregnant girl and her family insisted, and this was a decision in which the wishes of his first wife could play no part. She would certainly have left him had she had parents living, but she had only brothers. Similarly, the head of household number 9 had taken a second wife, whereupon his first wife had left him. The eldest son of household number 14 was unmarried owing to his addiction to opium at an early age. In household number 15, the youngest son had taken a first wife who had died in childbirth, he took up opium and remarried, but his second wife left him because of his addiction. Household number 17 was polygynous, but the wives were maintained in two separate establishments—a Thai rather than Hmong pattern! Household number 20 had, in fact, taken a second wife, the daughter of household number 3, who as indicated in Chapter 5, had died leaving him with three children (resident in household number 3). The wife who remained with him was his first. The wife of household number 23, an impoverished addict, had left him for a Thai trader, but he had no plans for a second marriage.

Of simultaneous polygynous marriages, therefore, there had only been four actual cases, one of which was imposed on the husband despite his strong resistance and reluctance. In two of the remaining cases, the wives had failed to put up with the arrangement to such an extent that one had left her husband, while the other husband had to maintain his wives in two separate establishments. In the fourth case, a love-match which was not properly celebrated, the second wife had died, and this really constituted the only genuine case of polygyny in the village. When questioned about this, the villagers would say that they disliked polygyny in general; usually the wives did not get on together, one had to be wealthy to acquire a second wife, and would normally only do so in case of infertility, when all other therapeutic or shamanic methods had failed to result in the birth of a child. In this the White Hmong seem to differ from the Green Hmong, who have much higher rates of polygyny (cf. Lemoine 1972a; Geddes 1976).

APPENDIX 4
Nomya Opium Yields (after the 1982 harvest)

Household	Opium Yields by Joi (1.6 kg)	Mature Labour Capacity
1	10	6*
2	8	5
3	5	4
4	4	1
5	10	2
6	8	7*
7	15	7*
8	4	2
9	15	7*
10	20	6
11	7	2
12	10	5*
13	9	4*
14	4	2
15	2	2
16	8	5*
17	6	2
18	6	6
19	2	3
20	4	5
21	5	2
22	8	2
23	5	4
Total	175	91

Notes

It is notoriously difficult to collect accurate statistics of shifting opium poppy cultivation. However, I place considerable reliance on these figures, which were intensively cross-checked. One household was not covered, besides the two newcomers to the village, and of the twenty-three which were, the order in which they were placed does not necessarily correspond to the order of households elsewhere in the book owing to reasons of confidentiality. However, I have placed the number of active over fourteen year olds in each household to the right of the figures. At the same time, the labour of children younger than fourteen can be very valuable in opium production, while very much depends on simply how hard a family works. I have indicated asterisks (*) for households where opium fields were held separately by different members of the household to a significant extent, but the figures given relate to the yields for the whole household. The average yield per household is, therefore, approximately 7½ joi (or 12 kg p.a.). Although this figure is generalizable, the relationship between yield and labour capacity does not seem to be a constant one, since overall adults produced under 2 joi each, while we can see that a household of one active adult was able to produce 4 joi relatively unaided.

The revenues derived from opium production should not be overestimated. If we look at household number 2, for example, which sold 5 of the 8 joi of opium produced at a total price of 16,000 baht, we must divide this by eight for each of the members of the household to arrive at a yearly income figure of 2,000 baht, or 167 baht per month. This is only just over the official poverty line in Thailand, defined at 1,981 baht per person in the rural areas in 1979 (Meesuk 1979).

Even if we add to this income that derived from the sale of vegetables (lettuce and beans) on two separate occasions during August and October in the local Chinese market, which fetched 2,300 baht (about two-thirds of what should have been paid), the total still decreases when one considers that while this was the only cash income received during the year, 800 baht was spent on ready-made cloth for clothing, five sacks of rice were bought from Chiangmai at 500 baht each (2,500 baht), and a further 500 baht spent employing four Karen labourers while preparing the fields. One also has to remember the many incidental expenses on medicines, exercise books for those children attending the project school, and so on. Pigs could have been sold for between 2,000 and 5,000 baht each, but none were as they are needed for health, both in the form of protein and for the curing rituals at which they are slaughtered. Weed-killer sold for 80 baht per liquid can (and several cans were invested in since the grasses proved so ferocious), and many villagers bought 15–20 large sacks of fertilizer at 350 baht each (although the PWD project in the larger settlement distributed one sack to each household there at the lower price of 300 baht per sack, repayable the following year). One villager (Table 5) grew no rice, but bought all from Chiangmai in return for good sales of 3 joi of opium.

I did not gather comprehensive data on maize yields, but most families grew enough to fill a large-size granary, while a family with sixty chickens and fourteen pigs (only four fully sized ones) used two buckets of corn a day feeding them, mixed with swill. Corn was mixed with rice to eat when rice stocks were very low, and also mixed into flour cakes towards the end of the year, as a sweetmeat. Where maize yields were plenteous, however, rice yields were not, as we have seen in the need to buy rice from Chiangmai or from local Karen at much lower prices.

The highest rice yield reported was 200 poom (about 100 bushels) by two families, one of which, with seven members to feed, still spent as much as 500 baht on rice from outside the village. Yields generally varied from 50 to 200 poom. At the same time, consumption figures were high (a family of two adults and two small children ate 2 bip of rice every thirteen days, or a little over 3 litres a day, while the household with twenty-five members, many young children, consumed the same amount in two days), and not all opium produced could be sold, as a large proportion is consumed internally, either by addicts or in hospitality to shamans or visiting guests who might themselves be purchasing opium, or in repayment of debt. Until very recently, the headman's family had had to pay 3–4 joi of opium a year to a Chinese trader in return for rice each year. A very typical, *direct*, sale of opium took place on the return of the defector from the CPT, whose elder brother instructed a younger clan member of the village to sell 2 *pob* (joi) to a Chinese trader in the mining community for 5,000 baht, half to be paid in cash and half in rice, which gave him five gunny bags, or about 30 bip, of rice. The major expenditures of villagers were, therefore, medicine, weed-killer, fertilizer, cloth, and, above all, rice.

APPENDIX 5
The Chao Fa Script

Consonants (*Suab qauv nkawm*)

vau	nrau	fau
ntsau	tsau	phau
lau	dau	dhau
hau	thau	plau
nkau	ntxau	rhau
rau	nphau	nplhau
mlau	hmlau	gau
nau	nqau	nqhau
qhau	hnyau	hmau
yau	ncau	sau

hlau	zau	ntxhau
xau	nau	nyau
cau	nrhau	txau
mau	txhau	qau
nchau	ntsau	au
hnau	khau	ntau
chau	xyau	tau
plhau	tshau	pau
nthau	nplau	nkhau

Vowels (*yub qauv nkawm*)

keeg	kuag	kawq	keg	kog	kwg
kees	kuas	kaws	kes	kos	kws
kee	kua	kaw	ke	ko	kw
keev	kuav	kawv	kev	kov	kwv
keej	kuaj	kawj	kej	koj	kwj
keem	kuam	kawm	kem	kom	kwm
keeb	kuab	kawb	keb	kob	kwb

kag	kiag	kug	kaig	koog	kig
kas	kias	kus	kais	koos	kis
ka	kia	ku	kai	koo	ki
kav	kiav	kuv	kaiv	koov	kiv
kaj	kiaj	kuj	kaij	kooj	kij
kam	kiam	kum	kaim	koom	kim
kab	kiab	kus	kaib	koob	kib

Numbers (*kobcub*)

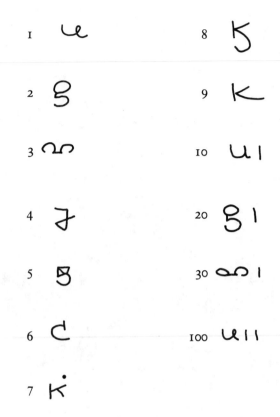

1		8	
2		9	
3		10	
4		20	
5		30	
6		100	
7			

Notes

The printed version I have differs a little from this. Since Hmong is monosyllabic, as in the Romanized Phonetic Alphabet (RPA), only two characters are needed to indicate the initial consonant (or consonant cluster, e.g., *npl-*), and the final, open-ended vowel (or end-nasalized vowel, as in the *oo* of Hmoob, for Hmong). The consonants are all clearly distinguished in this script, although the triple variants which are used for each character bear no relation to the consonants they indicate. Unfortunately, no consistent system of indicating tones seems to have been worked out, so that each vowel has seven variants for each of its tones. This means that there are seven times as many vowels as there need to be (ninety-eight rather than fourteen), while the lack of consistency between the sets of triple characters and the consonants they indicate makes it difficult to learn. While a tonal system for the vowels could easily be established, in a way its absence only shows that tone as meaning is more inherent to the native speaker of a tonal language than it is to foreign linguists; there was evidently an absence of awareness that vowels with different tonal values were not different animals altogether. For the quite arbitrary assignment of different consonantal values to triplets of similar characters, however, there is no such excuse, and it merely makes the characters for consonants hard to memorize for no very good reason.

The major difference between the RPA and the Hmong script is that final vowels are placed *before* the consonants, rather than after them as in RPA; and there are obvious precedents for this in the way certain vowels are written in Thai, Laotian, or other Pallava-derived scripts.

While there is no need to deny the revelatory nature of its emergence, the alphabet, like the *koom haum* movement of which it is a part, is quite clearly syncretic, and many points of resemblance can be noted, for example, between roman and Thai characters (while it has been suggested that some characters are derived from Russian markings on Vietnamese artillery). The music associated with the movement, too, influenced as it was by the American folk-influenced pop culture of pre-revolutionary Laotian student culture, shows odd transitions between pentatonic and a diatonic scale. (For an earlier version of this script, see Lemoine 1972c.)

APPENDIX 6
The Story of Yaj Xeeb Xeeb

This is a story about we Hmong a long time ago. I don't know if it is true or not. But Hmong everywhere keep on telling the same story. Yaj Xeeb Xeeb, Pov Sij Txam, and Yawg Laus Kiam, were the three who ruled (supported) the earth. Yaj Xeeb Xeeb was the head of the three. Faj Tim lived separately, in a different place. At that time, men and spirits could communicate with each other. One day Huaj Leej Pov Sij Txam came to Yaj Xeeb Xeeb and said, 'Great One, my wife has just given birth to a baby, but there is no food to eat. I want you to predict if this is a good day to fish or not.' Yaj Xeeb Xeeb replied, 'Today is a good day, because it is the fifth day. Go down to the river and you will find three fishes. Two are big, and one is small.' Pov Sij Txam went and found the three fishes as he had said. He killed them and brought them home.

The next day the Old Dragon (*Zaj Laug*) sent his young daughter to ask Yaj Xeeb Xeeb to his youngest son's wedding. Yaj Xeeb Xeeb said, 'Never before has the Old Dragon asked anyone to go down into his world. Today he has invited me to go and drink with him. I will go down there and see what there is good, and what is bad.' And he said to Pov Sij Txam and Yawg Laus Kiam, 'You two stay here. I am going to the spirit world, to see what is good there, and what is bad.'

He went into the Dragon's Lake and he entered the water and started to sink deeper and deeper under the water until his head was under the water and then he found that there were people living there in villages, just like on this earth. They were drinking and eating and they offered him some liquor to drink. Then the Old Dragon started to say, 'Yaj Xeeb Xeeb, you are the one who knows all things, but why did you let Pov Sij Txam shoot three of my guests?' Yaj Xeeb Xeeb said, 'I did not tell him to shoot your guests. I only told him to kill fish.' Then the Old Dragon said, 'If that is so, then let us compete to see who can tell how much rain will fall and how much the sun will shine this year.' Yaj Xeeb Xeeb said, 'This year is to be a good one. There will be nine parts of rain, and only three of light. So the harvest will be good, people will have lots to eat and will not suffer.' It was the Old Dragon who controlled the rains and waters, and he saw that Yaj Xeeb Xeeb already knew, so he said, 'This year there will be only three parts of rain, and nine of the sun.' Yaj Xeeb Xeeb replied, 'Unless you change your mind, everything will happen as I said it would.'

When he had finished eating, Yaj Xeeb Xeeb left and returned home. To Pov Sij Txam he said, 'You killed those three fish, and the Old Dragon says they were his guests, so that he is very angry now. I know he will change his mind and in the seventh month let it rain only three parts to nine parts of light. Everybody must beware since only half of the world will be able to survive. So go and take the head of the Old Dragon.'

But the Old Dragon had overheard him, and went to ask Faj Tim to help him. Faj Tim was sleeping, and as if in a dream, yet not in a dream, he saw the Old Dragon begging him to help him. And the Old Dragon said to him, 'Please help me. I was drinking at my son's wedding and invited Yaj Xeeb Xeeb to come and we competed to see whether the harvests would be good or not. I changed my mind and was going to let it rain only three parts and the sun shine for nine. Yaj Xeeb Xeeb was angry and is going to take my head. If you help me, I will give you one pillow of silver and one pillow of gold.' And when Faj Tim woke up, there he saw a pillow of gold and another pillow of silver beside him.

In those days it was possible for people to go up to Heaven and come down again, so Faj Tim said to his wife, 'Today I am going up to Heaven at 8 a.m. If Pov Sij Txam comes after 8 a.m., then try to stop him leaving; if he comes before then, we will both try to stop him from going on.'

And at 6 a.m. Pov Sij Txam arrived, riding on horseback, with the (jingling) sound of bells. Faj Tim went out to greet him. 'Has something happened?' he asked. Pov Sij Txam said, 'Yes, there is bad news.' Faj Tim urged him, 'Come inside and rest for awhile. Then you can go on. Why, we have not met since I sent you to rule the world.'

And in this way Faj Tim tried to stop Pov Sij Txam from leaving, killing a chicken and a pig to feed him. After that he said, 'I would like to have a spear of silver and gold, and I have heard you can make one. Why don't you make one for me?' Pov Sij Txam said, 'I will make one later, but now I have to leave,' but Faj Tim said, 'You need not leave now. When I say you can leave, then you can go.' So Pov Sij Txam agreed, and he made a fire and worked all night, and went to sleep.

Faj Tim said, 'Why are you taking such deep breaths?' Then he replied, 'It is because I am carrying the head of the Old Dragon (in spirit).' Faj Tim said, 'I don't believe it.' 'If you don't believe it,' he replied, 'go outside and see.' So Faj Tim went out, and at the gate he could see the Old Dragon with shining eyes and nine horns and eight fangs.

Faj Tim was afraid, and came back into the house, but Pov Sij Txam told Faj Tim, 'Don't be afraid, even if the Old Dragon has come to ask for your help. You're not the one who governs; we are the ones who govern, and you only supervise us. We will look after everything. The Old Dragon has done wrong to make people hungry and nine parts of the earth dry, so whatever happens, we will help you.'

At that time Faj Tim had only recently accepted his duty. Kob Lwj Txwv and Xeeb Kob Xwb were the first two. Huaj Leej Pov Sij Txam and Yaj Xeeb Xeeb were the two who came later. Kob Lwj Txwv and Xeeb Kob Xwb were the two that fought the war.

The Old Dragon had gone beneath the Heavens (i.e., his spirit had gone, since he was dead). The Dragon went to tell Xeeb Kob Xwb to tell Faj Tim what had happened to him.

So Xeeb Kob Xwb went to see Faj Tim. Faj Tim was afraid, and said, 'You have already died. Why are you here?' And Xeeb Kob Xwb replied, 'I have not died. I have only been resting. I have been helping you keep the peace in the land all this time. In heaven there are problems. The Old Dragon says that on earth he had some disagreement with Yaj Xeeb Xeeb and Pov Sij Txam and they cut off his head so that now he has no head, and he asked you, the ruler of earth, to help but you did not help him. So now he is asking you to come to be questioned.' Faj Tim started to cry. 'Don't cry,' said Xeeb Kob Xwb, 'I am here. You need not be afraid.' Faj Tim said, 'But heaven is not a physical world, it is a spiritual world, and if I go there I shall never be able to come back again, and I shall never see my family again.' Xeeb Kob Xwb said, 'If you are careful you will be able to return. Put your head under my arm, and I'll take you there.' And he did.

As they reached heaven, he heard the sound of voices crying. 'Why do they cry in heaven, just like on earth?' he asked. 'They are all those who killed themselves by shooting, or eating poison, or hanging themselves from trees,' said Xeeb Kob Xwb. 'They cannot go (further), but all those who died of illness can go through.'

And as they went on their way he heard more crying, and Faj Tim was very afraid, but Xeeb Kob Xwb said, 'Don't be afraid. We have our own problems, as they have theirs.' And finally they got there.

'Oh, Faj Tim,' said the Old Dragon, 'I asked you to help me but you did not, and now, look, I don't have my head. What will you say?' Faj Tim said, 'I tried hard to help you. I tried to kill a chicken and pig and cook them very slowly so that when he had finished you would be so far away he could not catch up with you. But he hadn't done it physically, he had done it in spirit, and that was why I could not help you.'

'You did not bring the pillows with you when you came here. When you knew you couldn't help, you should have returned the pillows of silver and gold,' said the Old Dragon.

'Now I came with Xeeb Kob Xwb and we left that silver and gold back on earth. But don't worry (impl. you shall have it back),' Faj Tim told him.

But he had not told the Old Dragon how he was going to get it back to him. So he spent a whole day returning, but in the morning still found himself in the Old Dragon's house. After nine days returning he was still there in the Old Dragon's house and never got home, so he went to ask Xeeb Kob Xwb about it: 'We finished the problems already,' he said, 'so why can't I go home?' And Xeeb Kob Xwb said, 'You forgot to bless him.' 'How shall I bless him?' asked Faj Tim. 'Tell him you are sorry you could not help him and now he has no head, but that he is still alive and still rules the earth. Tell him that after you have gone back, when it rains and the sun shines at the same time, he (will) rise up as a rainbow, and we will all say that Tsib Looj has risen up. In this way we will be greeting him so he is not forgotten.' And after that he was able to return.

On their way, Xeeb Kob Xwb asked Faj Tim if he wanted to look at the earth from where they were. Not knowing that they had tied up Tuam Looj Kob and Kob Lwj Txwv because they had killed so many people in war so that Ntxwj Nyug had had them tied up, Faj Tim replied that he would. If he hadn't wanted to look, he would have arrived home. But because he did, Xeeb Kob Xwb took him to see. And as they opened the courtyard gate he saw Tuam Looj Kob and Kob Lwj Txwv in chains. They were very happy to see him, and said, 'We are glad you have come here. Now you can take our place and we can go. It's all because of you we're tied up. You wanted to take over the country and we contested with you and you won so that we had to come here and be punished. And we have given up (have disgust of) money to buy anything to eat, so now you stay here instead of us and let us go.' So Faj Tim began to cry. But Xeeb Kob Xwb said, 'Don't worry. Ask them if they need money, then you can go and borrow three big baskets of silver and three big baskets of gold from the Chinese couple for them. Then you can go back to earth.'

So he went to ask them, and they said, 'Since we are earthly people, if you can find some money for us, then we will let you go.' And he went to borrow three big baskets of silver and three big baskets of gold to give them to buy some food. Then they went on. 'Do you still want to see the world?' asked Xeeb Kob Xwb. If he had refused he would have returned home, but he agreed, and so they went. And as they opened the gate, he saw that his body upon the earth was already dead. And he saw his wife chasing the flies away from his body with a fan. So he started to cry. But Xeeb Kob Xwb said, 'Don't cry. You will be able to return. It is because you have been here so long that your body has died.' And so they went on until they reached the edge of the river. Then Xeeb Kob Xwb said, 'You stay here. I am going back.' And as he waited he heard the roar of thunder and the skies darken and the rhinoceros bellowing at the far waterside, so he jumped up on its back, and it began to kick. And as it kicked, the lid of the coffin opened up and his wife, going to look, saw that he had risen up. But he was unable to speak, and after three years still could not say a single word.

They took him all over the country to people to try and trick him into speaking, but could not find one who was able to turn over his licence for life. Finally, they came to Tuam Som Yej. And Tuam Som Yej said, 'When he went to Heaven, he went for such a long time with his *plig* (soul) that his body died. There he met his two overlords, and they asked him to take their place, but he did not. He borrowed silver and gold for them from the Chinese which he had to pay back. But for silver, you should not have found real silver, and for gold, you should not have found real gold. You should only have found the things that people in the spirit-world need. Heavenly money is earthly tears, horses and cattle in the world of spirits are love songs in the world of men. So do not look for real (earthly) silver or gold. For silver you must find white paper, for gold you must find golden paper. Make them into flowers, and burn three basketsful of each of them. Then let them pass along the river, calling the spirits and asking them to let Faj Tim speak again. And then Faj Tim will be able to speak again.'

They went back home, and did all that Tuam Som Yej had told them. And then Faj Tim was able to speak again.

It is true that in ancient times (at the beginning of things), Hmong were (really) able to go up to heaven (the spirit world), and talk to the *dab* (spirits). They could communicate with the spirits. And now they know (the difference between) what is spirit silver, and what is earthly silver (money).

Suav Yaj,
Nomya, August 1981

The latter part of the story refers to the practice which the Hmong share with the Chinese, of burning votive paper while performing sacrifices and on other ritual occasions, which is believed to become money in the Otherworld. The Hmong term for the rainbow, *zaj sawv*, literally means a 'dragon rising'.

APPENDIX 7
The Panthays or Hui-hui

They furnished the late Sir Edward Sladen with the following account of themselves in 1869:

The Chief Queen of the Emperor Tanwan adopted a child and called him Anlaushan. In time the child developed into a man of extraordinary comeliness and wonderful intellect.

The Queen was enamoured and the adopted son became her paramour. Anlaushan soon rose to distinction. His abilities were of the highest order and raised him at once to fame and influence. The queenly passion was not disclosed; but suspicion had been sufficiently roused to make it prudent on the Queen's part to get rid of her lover and defeat all signs of illicit intercourse.

Anlaushan was accordingly accused of being privy to a conspiracy to dethrone the Emperor. The influence of the Queen prevailed to obtain a conviction and his favourite was banished from the royal capital.

But the injustice of his accusation and a sense of wrong roused Anlaushan to action and induced him to become in reality a leader of rebellion. He lost no time in collecting a large force with which he was able to make head against the Government and successfully encounter the troops of the Emperor. In time he had approached within a league of the capital and city and palace were alike threatened.

The Emperor Tanwan in this emergency adopted the suggestion of his Vizier Kanseree and despatched a mission to Soeyoogwet and implored foreign aid. A force of 3,000 men was sent under the command and guidance of three learned teachers, who arrived in due time at Tanwan's capital. By their aid Anlaushan was defeated and eventually captured.

The rebellion was at an end and the foreign contingent left China to return to its own country. Here, however, a difficulty arose. Their rulers refused them admittance and alleged as a cause for doing so that it was against the constitution of the country to receive back men who had come into combat with pork-eating infidels. They had herded in fact with pigs and infidels, and could no longer be regarded as unpolluted subjects, or as fit members of a society which held pork in religious detestation.

They returned therefore to China and became permanent sojourners in a foreign land. They are the original stock from which Mahomedanism has sprung up in China, in various communities and under several denominations.

J. Anderson (1878) identifies Tanwan with (T'ang) Huan (now more commonly written Yuan) Tsung, against whom Ngan Lo-shan rebelled. The *ng* are letters of supererogation frequently omitted. The next emperor of the T'ang (not Tung as J. Anderson writes it) was Su Tsung, who acceded in AD 756 and was rescued from his difficulties by the arrival of an embassy from the Khalif Abu Jafar Al Mansur, the founder of Baghdad, accompanied by auxiliary troops who were joined by Ouighour Tartars and other forces from the west.

This account of their origin seems farfetched when we remember Kublai Khan's conquest of Tali 500 years later and the existence of the Tungamis in the north.

Nevertheless, the Hui-hui in the days of their independence believed it, for they sent the following letter to Sladen (quoted in Scott and Hardiman 1900, pp. 607–8):

The Panthays send greeting to their friends.

When Lanlu and other Kachins came to Momien we conversed with them freely and were extremely happy to learn that three hundred foreigners had arrived in Bhamo.

Being of the same belief as yourselves, we know your willingness to help and assist us.

We are the descendants of three thousand men of the Lerroo country, who, being unable to return to their native land, settled down in China, where we have been upwards of a thousand years. Some ten years ago the Chinese Government became so intolerably oppressive that by God's help Tuhin-shee of the Tu race (that is to say, his *hsing*, or surname) was commanded to separate the good from the wicked and obtained possession of the western provinces of the Chinese Empire. . . .

Bibliography

Abadie, M., *Les Races du Haut-Tonkin de Phong-Tho a Lang-Son*, Paris, Ancienne Maison Challemel, 1924.

Abhay, Thao Nhouy and Nhinn, P. S. (trans.), *Sinsay, chef d'oeuvre de la litterature lao*, Bangkok, Liang Chong Charoen Press, 1965.

Ajchenbaum, Y. and Hassoun, J-P., *Histoires d'insertion des Groupes Familiaux Hmong Refugiés en France*, Paris, Association pour le Developpement de la Recherche et de l'Experimentation en Sciences Humaines, 1980.

Allton, I., 'La guerre du Fou: essai d'Interpretation', unpublished MA thesis, CeDRASEMI, Paris, 1978.

Althusser, L., 'Ideology and Ideological State Apparatus: Notes towards an Investigation', in *Lenin and Philosophy, and Other Essays*, London, NLF Books, 1971.

Anderson, J., *A. R. Margary, The Journey of, From Shanghai to Bhamo, and Back to Manwyne*, London, Macmillan and Co., 1878.

Anderson, P., *Lineages of the Absolutist State*, London, Verso Editions, 1974.

Archaimbault, C., 'Religious Structures in Laos', *Journal of the Siam Society*, 52, 1964.

Banton, M., 'Rational Choice: A Theory of Racial and Ethnic Relations', SSRC Working Papers on Ethnic Relations, No. 8, SSRC, 1977.

Barney, G. L., 'Christianity and Innovation in Meo Culture: A Case Study in Missionization', unpublished MA thesis, University of Minnesota, 1957.

Benedict, P., *Austro-Thai Language and Culture*, New Haven, Human Relations Area Files Press, 1975.

Benjamin, W., in H. Arendt (ed.), *Illuminations*, London, Jonathan Cape, 1979.

Berlin, I., *Historical Inevitability*, London, Oxford University Press, 1954.

Bernatzik, H. A., *Akha and Miao: Problems of Applied Ethnography in Further India*, 1947, New Haven, trans. Human Relations Area Files Press, 1970.

Bernstein, B., 'Social Class, Language and Socialization', in *Class, Codes and Control*, Vol. I, London, Routledge and Kegan Paul, 1971.

Bertrais, Y., *Dictionnaire Hmong (Meo Blanc)–Francais*, Laos, Mission Catholique, Vientiane, 1964.

_____, *The Traditional Marriage among the White Hmong of Thailand and Laos*, Chiangmai, Hmong Centre, 1978.

Betts, G. E., 'Social Life of the Miao tsi', *Journal of the Royal Asiatic Society, North China Branch*, Nos. 1–21, Shanghai, 1899–1900.

Binney, G. A., 'The Social and Economic Organization of Two White Meo Communities in Northern Thailand', unpublished Ph. D. thesis, Advanced Research Projects Agency, Department of Defense, Washington, DC, 1971.

Bliatou, B., *Hmong Sudden Unexpected Nocturnal Death Syndrome: A Cultural Study*, Portland, Oregon, Sparkle Publishing Enterprises, 1982.

Bloch, E., *Geist de Utopie*, Munich and Leipzig, Suhrkamp Verlag, 1918.

Bloch, M., 'The Past and the Present in the Present', *MAN*, new series 18, 1977.

Bourotte, B., 'Marriages et funerailles chez les Meo blanc de la region de Nong-het (Tran Ninh)', *Bulletins et Travaux de l'Institut Indochinois pour l'Etude de l'Homme*, 6, 5, Hanoi, 1943.

Bunnag, T., 'Kabot phu mi bun phak Isan', *Sangkhomsat Parithat* (Social Science Review), 5, 1, 1967 (Thai).

Burridge, K., *New Heaven, New Earth*, Oxford, Blackwell, 1971.

Carr, E. H., *What is History?* London, Macmillan, 1961.

Cassirer, E., *The Philosophy of Symbolic Forms 1. Language*, New Haven, Yale University Press, 1953.

Chindasri, N., *The Religion of the Hmong Njua*, Bangkok, The Siam Society, 1976.

Christie, A., *Chinese Mythology*, London, Hamlyn, 1968.

Clark, G. W., *Kweichow and Yun-Yan Provinces*, London, 1894.

Clarke, S. R., *Among the Tribes in South-West China*, London, China Inland Mission, 1911.

Coedès, G., 'L'Origine du Cycle des Douze Animaux au Cambodge', *Toung Pao*, xxxi, 1935.

Cohen, A., *Custom and Politics in Urban Africa*, London, Routledge and Kegan Paul, 1969.

Cohen, P., 'The Politics of Economic Development in Northern Thailand 1967–1979', unpublished Ph. D. thesis, University of London, 1981.

Cohn, N., *The Pursuit of the Millenium*, London, Secker and Warburg, 1957.

Coleridge, S. T. C., in E. L. Griggs (ed.), *Letters*, Oxford, The Clarendon Press, 1956.

Collingwood, R. G., *The Idea of History*, Oxford, The Clarendon Press, 1946.

Condominas, G., 'A Few Remarks about Thai Political Systems', in G. Milner (ed.), *Natural Symbols in South East Asia*, London, School of Oriental and African Studies, 1978.

Cooper, R. G., 'The Tribal Minorities of Northern Thailand: Problems and Prospects', *Southeast Asian Affairs*, VI, 1979.

———, 'The Hmong of Laos: Economic Factors in Refugee Exodus and Return', paper presented to the International Conference on Hmong Studies, University of Minnesota, November 1983.

———, *Resource Scarcity and the H'mong Response: A Study of Settlement and Economy in Northern Thailand*, Singapore, Singapore University Press, 1984.

Croce, B., *History as the Story of Liberty*, London, Allen and Unwin, 1941.

Davenport, W., Olmsted, D., Mead, M., and Freed, R., *Report of the Ad Hoc Committee to Investigate the Controversy Concerning Anthropological Activities in Thailand*, American Anthropological Association, 1971.

Davies, H. R., *Yun-Nan: The Link between India and the Yangtze*, Cambridge, Cambridge University Press, 1909.

Davis, Fei-Ling, *Primitive Revolutionaries of China: A Study of Secret Societies in the Late Nineteenth Century*, London, Routledge and Kegan Paul, 1977.

Davis, R., 'Tolerance and Intolerance of Ambiguity in Northern Thai Myth and Ritual', *Ethnology*, 1, 1974.

De Beauclair, I., 'Ethnic Groups of South China', New Haven, Human Relations Area Files Press, 1956a.

———, 'Culture Traits of Non-Chinese Tribes in Kweichow Province, Southwest China', *Sinologica*, V, 1, 1956b.

———, 'A Miao Tribe of Southwest Kweichow and its Cultural Configuration', *Bulletin of the Institute of Ethnology*, 10, Academica Sinica, Nanking, 1960.

Derrida, D., *Of Grammatology*, Baltimore, Johns Hopkins University Press, 1974.

Dessaint, A., *Minorities of Southwest China*, New Haven, Human Relations Area Files Press, 1980.

Dhamaraso, Bhikhu and Virojano, Bhikhu, *The Meo*, Bangkok, 1973.

Dodds, E. R., *The Greeks and the Irrational*, London, 1928.

———, *The Ancient Concept of Progress, and Other Essays on Greek Literature and Belief*, Oxford, The Clarendon Press, 1973.

Dommen, A., *Conflict in Laos: The Politics of Neutralization*, New York, Praeger, 1971.

Dore, H., *Lao-Tse et le Taoism*, Shanghai, Impr. Mission Catholique, 1938.

Downer, G., 'Tone-Change and Tone-Shift in White Miao', *Bulletin of the School of Oriental and African Studies*, 30, 3, 1967.

Durkheim, E. and Mauss, M., in R. Needham (ed. and trans.), *Primitive Classification*, Chicago, University of Chicago Press, 1963.

Durrenberger, E. P., *Lisu Project: A Socio-Medical Study of the Lisu of Northern Thailand*, Chiangmai, Tribal Research Centre, 1970.

✓ Dyke, R., *Samuel Pollard, Missionary Supreme: A Centenary Tribute*, London, The Epsworth Press, 1964.

Eberhardt, W., *The Local Cultures of South and East China*, Leiden, Brill, 1968.

———, 'Marital Customs and Funerals of the Miaotze of Kweichow', in W. Eberhardt (ed.), *Studies in Chinese Folklore and Related Essays*, Bloomington, Indiana University Research Centre for the Language Sciences, 1970.

———, *Chinese Minorities, Yesterday and Today*, Belmont, California, Wadsworth Inc., 1982.

Edkins, J., 'Feng-shui', *Chinese Recorder and Missionary Journal*, March 1872.

Eitel, E. J., *Feng-shui, or, The Rudiments of Natural Science in China*, Hong Kong, Trubner and Co., 1873.

Endriquez, C., *Races of Burma* (Handbooks for the Indian Army), Calcutta, 1924.

Evans, G., *The Yellow Rainmakers*, London, Verso Editions, 1983.

Evans-Pritchard, E., *Witchcraft, Oracles and Magic among the Azande*, Oxford, The Clarendon Press, 1937.

———, *The Nuer: A Description of the Modes of Livelihood and Political Institutions of Nilotic People*, Oxford, The Clarendon Press, 1940.

———, 'Anthropology and History', in *Essays in Social Anthropology*, London, Routledge and Kegan Paul, 1961.

Feingold, D., *Proposal to Control Opium from the Golden Triangle and Terminate the Shan Opium Trade*. 'Statement' to the *Hearings* before the Subcommittee on Future Foreign Policy Research and Development, Committee on International Relations, House of Representatives, 94th Congress, 1st Session, 22 April 1975.

Ferlus, M., 'Le Recit Khamou de Chuang et ses Implications Historiques pour le Nord-Laos', *ASEMI*, x, 2–4, 1979.

Festinger, L., Riecken, H., and Schachter, S., *When Prophecy Fails*, New York, Harper, 1956.

Feuchtwang, S., *An Anthropological Analysis of Chinese Geomancy*, Vientiane, Vithagna, 1974.

———, 'Investigating Religion', in M. Bloch (ed.), *Marxist Analyses and Social Anthropology*, ASA Studies 3, London, Malaby Press, 1975.

Firth, R., 'Social Organization and Social Change', in *Essays on Social Organization*, London, University of London Press, 1964.

———, 'Rumour in a Primitive Society with a Note on the Theory of Cargo Cults', *Tikopia Ritual and Belief*, London, Allen and Unwin, 1967.

Fitzgerald, C. P., *China: A Short Cultural History*, London, The Cresset Press, 1935.

_____, *The Southern Expansion of the Chinese People*, London, Barrie and Jenkins, 1972.

Forbes, A., 'The Cin-Ho' (Yunnanese Chinese) Caravan Trade with North Thailand during the Late 19th and Early 20th Centuries', *Journal of Asian History*, 21, 2, 1987.

Fortes, M., *The Web of Kinship among the Tallensi*, London, Oxford University Press for the International African Institute, 1949.

Fortune, R., 'Yao Society: A Study of a Group of Primitives in China', *Lingnam Science Journal*, 18, 3, Canton, 1939.

Freedman, M., *Chinese Lineage and Society*, London, The Athlone Press, 1966.

_____, 'Chinese Geomancy', Presidential Address, Royal Anthropological Institute, 1968, *Journal of the Royal Anthropological Institute*, 1968 (1969): 5–15.

Fried, M., *The Notion of Tribe*, Columbia, 1975.

Friedman, J., 'Tribes, States and Transformations', in M. Bloch (ed.), *Marxist Analyses and Social Anthropology*, ASA Studies 3, London, Malaby Press, 1975.

Garrett, W. E., 'The Hmong of Laos: No Place to Run', *National Geographic Magazine*, 143, 1, 1974.

_____, 'Refuge from Terror', *National Geographic Magazine*, 157, 5, 1980.

Geddes, W. R., 'The Tribal Research Centre at Chiangmai', in P. Kunstadter (ed.), *Southeast Asian Tribes, Minorities and Nations*, Princeton, Princeton University Press, 1967.

_____, 'The Hill Tribes of Thailand', *SEATO Record*, IV, 6, 1975.

_____, *Migrants of the Mountains: The Cultural Ecology of the Blue Miao (Hmong Njua) of Thailand*, Oxford, The Clarendon Press, 1976.

_____, 'Research and the Tribal Research Centre', in J. McKinnon and W. Bhrukrasri (eds.), *Highlanders of Thailand*, Kuala Lumpur, Oxford University Press, 1983.

Geertz, C., *Agricultural Involution*, Berkeley, University of California Press, 1963.

Gesau, Alting von, 'Dialectics of Akhazan: The Interiorizations of a Perennial Minority Group', in J. McKinnon and W. Bhrukrasri (eds.), *Highlanders of Thailand*, Kuala Lumpur, Oxford University Press, 1983.

Giddens, A., *Capitalism and Modern Social Theory*, Cambridge, Cambridge University Press, 1971.

Gilhodes, C., 'Mythologie et Religion des Katchins (Birmanie)', *Anthropos*, 4, 1909.

Girling, J., *Thailand: Society and Politics*, Ithaca, Cornell University Press, 1981.

Godelier, M., *Rationality and Irrationality in Economics*, London, New Left Books, 1973.

Goldstein, B., 'Resolving Sexual Assault: Hmong and the American Legal System', paper presented to the International Conference on Hmong Studies, University of Minnesota, November 1983; published in E. Hendricks, B. Downing, and A. Deinard (eds.), *The Hmong in Transition*, Minneapolis, Center for Migration Studies of New York Inc. and the Southeast Asian Refugee Studies Program of the University of Minnesota, 1986.

Goody, J., *The Logic of Writing and the Organization of Society*, Cambridge, Cambridge University Press, 1986.

Graham, D. C., 'Ceremonies of the Ch'uan Miao', *Journal of the West China Border Research Society*, Chengtu, 9, 1937.

_____, *Songs and Stories of the Ch'uan Miao*, Miscellaneous Collections No. 123, Washington, DC, Smithsonian Institute, 1954.

_____, 'A Lolo Story: The Great God O-Li-Bi-Zih', *Journal of American Folklore*, 68, 1955.

Grandstaff, T., 'The Hmong, Opium and the Haw: Speculations on the Origin of their Association', *Journal of the Siam Society*, 62, 2, 1979.

Granet, M., *Chinese Civilization*, London, Kegan Paul, 1930.

_____, 'Right and Left in China', in R. Needham (ed.), *Right and Left: Essays in Dual Symbolic Classification*, Chicago, University of Chicago Press, 1973.

Grist, A., *Samuel Pollard: Pioneer Missionary in China*, Taipei, Cassell Reprint, 1971.

Haas, M., *Thai–English Student's Dictionary*, Stanford University Press, 1964.

Halpern, J., *Government, Politics and Social Structure in Laos: A Study of Trade and Innovation*, New Haven, Yale University Press, 1964.

Hanks, L., 'The Lahu Shi Hopoe: The Birth of a New Culture?', in L. Hanks and L. Sharp (eds.), *Ethnographic Notes on North Thailand*, Data Paper No. 58, Southeast Asia Programme, Cornell University, Ithaca, 1965.

Hanks, L., Hanks, J., Sharp, L., and Sharp, R., *Anthropological Survey of Hill Tribes in Northern Thailand: A Report on Tribal Peoples in Chiangmai Province, North of the Meo Kok River*, Siam Society Data Paper No. 1, 1964.

Hayes, E. H., *Sam Pollard of Yunnan*, Pioneer Series No. 8, London, The Pilgrim Press, 1928.

Heath, A., *Rational Choice and Social Exchange*, Cambridge, Cambridge University Press, 1976.

Heimbach, E., *White Hmong–English Dictionary*, Data Paper No. 75, Southeast Asia Programme, Cornell University, Ithaca, 1979.

Heimbach, M., *At Any Cost: The Story of Graham Ray Orpin*, London, Overseas Missionary Fellowship, 1976.

Herskovits, M., *The Economic Life of Primitive People*, 1940.

Hertz, R., *Death and the Right Hand*, R. and C. Needham (trans.), London, Cohen and West, 1960.

_____, 'The pre-eminence of the right hand: a study in religion and polarity' (1909), in R. Needham (ed.), *Right and Left: Essays on Dual Symbolic Classification*, Chicago, University of Chicago Press, 1973.

Hill, A., 'Familiar Strangers: The Yunnanese Chinese in Northern Thailand', unpublished Ph.D. thesis, University of Illinois at Urbana-Champagne, 1982.

Hinton, P., 'Introduction', in *Tribesmen and Peasants in North Thailand*, Chiangmai, Tribal Research Centre, 1969.

Hobsbawm, E. and Ranger, T. (eds.), *The Invention of Tradition*, Cambridge, Cambridge University Press, 1983.

Hosie, A., *On the Trail of the Opium Poppy*, London, George Philip and Son, 1914.

Hsiao-Tung, Fei, 'Ethnic Identification in China', *Social Sciences in China*, March 1980.

Hsu, F., *Under the Ancestors' Shadow: Chinese Culture and Personality*, London, Routledge and Kegan Paul, 1949.

Hudson, W., *The Marxist Philosophy of Ernst Bloch*, London, Macmillan, 1982.

Hudspeth, W., 'The Cult of the Door Amongst the Miao in South-West China', *Folk Lore*, 33, 1922.

_____, 'Among the Flowery Miao', *The Listener*, August 1932.

_____, 'The Hwa Miao Language', *Journal of the West China Branch of the Royal Asiatic Society*, 7, 1935.

_____, *Stone Gateway and the Flowery Miao*, London, The Cargate Press, 1937.

Ingram, J. C., *Economic Change in Thailand 1850–1970*, London, Oxford University Press, 1971.

Irvine, W., 'The Thai-Yuan "Madman" and the "Modernising, Developing Thai Nation" as Bounded Entities under Threat: A Study in the Replication of a Single Image', unpublished Ph.D. thesis, University of London, 1982.

Israeli, R., *Muslims in China: A Study in Cultural Confrontation*, Scandinavian Institute of Asian Studies Monograph No. 29, London, Curzon Press, 1978.

Izikowitz, K., *Lamet: Hill Peasants in French Indochina*, *Ethnologiska Studier*, 17, Goteburg, 1951.

Jones, M., 'In Place of the Poppy', *The Guardian*, 26 April 1977.

Josipovici, G., *The World and the Book: A Study of Modern Fiction*, London, Macmillan, 1971.

Keen, G., 'Zonal Development: The Basic Proposition', *SEATO Spectrum*, 2, 1973.

_____, 'Ecological Relationships in a Hmong (Meo) Economy', in P. Kunstadter, E. Chapman, and S. Sabhasri (eds.), *Farmers in the Forest: Economic Development and Marginal Agriculture in Northern Thailand*, Honolulu, East–West Center, University Press, Hawaii, 1978.

Kemp, E., 'The Highways and Byways of Kweichow', *Journal of the Royal Asiatic Society (West China Branch)*, 52, 1921.

Kempis, Thomas à, *The Imitation of Christ*, London, J. M. Dent and Son, 1928.

Kendall, R. Elliott (ed.), *Eyes of the Earth: The Diary of Samuel Pollard*, London, The Cargate Press, 1954.

Keyes, C., 'Buddhism and National Integration in Thailand', *Journal of Asian Studies*, xxx, 3, 1971.

_____, *The Golden Peninsula: Culture and Adaptation in Mainland Southeast Asia*, London, Macmillan, 1977.

_____, (ed.), *Ethnic Adaptation and Identity: The Karen on the Thai Frontier with Burma*, Philadelphia, Institute for the Study of Human Issues, 1979.

Khinthitsa, 'Nuns, Mediums and Prostitutes in Chiangmai: A Study of Some Marginal Categories of Women in Chiangmai', Occasional Paper No. 1, Centre of South-East Asian Studies, University of Kent at Canterbury, 1983.

Koch, A., 'Collective Protest Movements in Siam between 1850–1930: The Significance of Ideology', unpublished MA thesis, Institute of Social Studies, The Hague, 1981.

Koestler, A., 'The Limits of Confirmation', in *The Act of Creation*, London, Hutchison, 1964.

Kraisri Nimmanhaeminda, 'Put Vegetables into Baskets, People into Towns', in L. Hanks, J. Hanks, and J. Sharp (eds.), *Ethnographic Notes on Northern Thailand*, Ithaca, Cornell University Press, 1965.

Kuhn, I., *Ascent to the Tribes: Pioneering in North Thailand*, London, Overseas Missionary Fellowship, 1956.

Kunstadter, P. (ed.), *Southeast Asian Tribes, Minorities and Nations*, Princeton, Princeton University Press, 1967.

_____, 'Ethnic Group, Category and Identity: Karen in Northern Thailand', in C. Keyes, *Ethnic Adaptation and Identity: The Karen on the Thai Frontier with Burma*, Philadelphia, Institute for the Study of Human Issues, 1979.

_____, 'Highland Populations in Northern Thailand', in J. McKinnon and W. Bhrukrasri (eds.), *Highlanders of Thailand*, Kuala Lumpur, Oxford University Press, 1983.

Lam Tam, 'A survey of the Meo', *Vietnamese Studies*, Hanoi, Ethnographic Data, 1972.

Latourette, K. S., *A History of Christian Missions in China*, Taipei, 1929; reprinted New York, Russell and Russell, 1967.

Laurence, P., 'The Fugitive Years: Cosmic Space and Time in Melanesian Cargoism and Medieval European Chiliasm' in R. Wallis (ed.), *Millenialism and Charisma*, The Queen's University of Belfast, 1982.

Leach, E. R., *Political Systems of Highland Burma: A Study of Kachin Social Structures*, London, The Athlone Press, 1954.

———, 'The Frontiers of Burma', *Comparative Studies in Society and History*, 3, 1960–1.

———, 'Virgin Birth', *Proceedings of the Royal Anthropological Institute*, 1966: 39–49.

Lebar, F. M., Hickey, G. C., and Musgrave, J. K., *Ethnic Groups of Mainland Southeast Asia*, New Haven, Human Relations Area Files Press, 1964.

Lee Chee-Boon, 'Local History, Social Organization and Warfare of the Yao', in R. F. Fortune (ed.), *Lingnam Science Journal*, 18, 3, Canton, 1939.

Lee, Gar Yia, 'The Effects of Development Measures on the Socio-Economy of the White Hmong', unpublished PhD thesis, University of Sydney, 1981.

———, 'Culture and Adaptation: Hmong Refugees in Australia', paper presented to the International Conference on Hmong Studies, University of Minnesota, November 1983; published in E. Hendricks, B. Downing, and A. Deinard (eds.) *The Hmong in Transition*, Minneapolis, Center for Migration Studies of New York Inc. and the Southeast Asian Refugee Studies Program of the University of Minnesota, 1986.

Lehman, F. K., *The Structure of Chin Society*, Urbana, University of Illinois Press, 1963.

———, 'Ethnic Categories in Burma and the Theory of Social Systems', in P. Kunstadter (ed.), *Southeast Asian Tribes, Minorities and Nations*, Princeton, Princeton University Press, 1967a.

———, 'Burma: Kayah Society as a Function of the Shan–Burma–Karen Context', in J. Steward (ed.), *Contemporary Change in Traditional Society*, Vol. II, Urbana, University of Illinois Press, 1967b.

———, 'Who are the Karen, and If So, Why? Karen Ethnohistory and a Formal Theory of Ethnicity', in C. Keyes (ed.), *Ethnic Adaptation and Identity: The Karen on the Thai Frontier with Burma*, Philadelphia, Institute for the Study of Human Issues, 1979.

Lemoine, J., *Un Village Hmong Vert du Haut Laos: Milieu Technique et Organisation Sociale*, Paris, Centre National de la Recherche Scientifique, 1972a.

———, 'L'Initiation du Mort chez les Hmong', *L'Homme*, XII, 1–3, 1972b.

———, 'Les Ecritures du Hmong', *Bulletin des Amis du Royaume Lao*, 7–8, 1972c.

———, *Yao Ceremonial Paintings*, Bangkok, White Lotus Co., 1982a.

———, 'La Malediction des Hmong en Exil: Les Reves qui Tuent', *Actuelle*, November 1982b.

———, 'Yao Religion and Society', in J. McKinnon and W. Bhrukrasri (eds.), *Highlanders of Thailand*, Kuala Lumpur, Oxford University Press, 1983.

———, 'Shamanism in the Context of Hmong Resettlement', paper presented to the International Conference on Hmong Studies, University of Minnesota, November 1983b; published in E. Hendricks, B. Downing, and A. Deinard (eds.), *The Hmong in Transition*, Minneapolis, Center for Migration Studies of New York Inc. and the Southeast Asian Refugee Studies Program of the University of Minnesota, 1986.

Levi-Strauss, C., *Tristes Tropiques*, Paris, Librarie Plon, 1955.

———, 'Social Structure', in *Structural Anthropology*, Harmondsworth, Penguin University Books, 1963a.

———, 'The Structural Study of Myth', in *Structural Anthropology*, Harmondsworth, Penguin University Books, 1963b.

———, 'Race and History', in *Structural Anthropology II*, London, Allen Lane, 1977.

Li Chi, *The Formation of the Chinese People: An Anthropological Inquiry*, Cambridge, Mass., Harvard University Press, 1928.

Lin Yueh-hua, 'The Miao-Man Peoples of Kweichow', *Harvard Journal of Asiatic Studies*, 5, 3–4, 1940.

———, 'Social Life of the Aboriginal Groups in and around Yunnan', *Journal of the West China Branch of the Royal Asiatic Society*, 1944.

Liu Chung-shee, 'The Dog-Ancestor Story of the Aboriginal Tribes of Southern China', *Journal of the Royal Anthropological Institute*, 62, 1932.

Loewe, M., *Chinese Ideas of Life and Death: Faith, Myth and Reason in the Han Period*, London, Allen and Unwin, 1982.

Lombard-Salmon, C., *Un Exemple d'Acculturation Chinoise: La Province du Gui Zhou au XVIIIᵉ Siecle*, Paris, École Francaise d'Éxtrême-Orient, 1972.

Lunet De La Jonquière, L., *Ethnographie du Tonkin Septentrional*, Ernest Leroux, Paris, 1906.

Lyman, T. A., *English/Meo Pocket Dictionary*, Bangkok, Goethe Institute, 1970.

MacAleavy, H., *Black Flags of North Tonkin*, London, Allen and Unwin, 1968.

McCarthy, J., *Surveying and Exploring in Siam*, London, John Murray, 1900.

McCoy, A. W., *The Politics of Heroin in Southeast Asia*, New York, Harper and Row, 1972.

McKinnon, J. and Bhrukrasri, W. (eds.), *Highlanders of Thailand*, Kuala Lumpur, Oxford University Press, 1983.

Mandorff, J., 'A Report on the Establishment of a Tribal Research Centre in Chiangmai', in P. Kunstadter (ed.), *Southeast Asian Tribes, Minorities and Nations*, Princeton, Princeton University Press, 1967a.

———, 'The Hill Tribe Programme of the Public Welfare Department', in P. Kunstadter (ed.), *Southeast Asian Tribes, Minorities and Nations*, Princeton, Princeton University Press, 1967b.

Mangrai, Saimong, *The Padaeng Chronicle and the Jengtung State Chronicle Translated*, Ann Arbor, University of Michigan Press, 1981.

Marx, K., *Early Writings*, London, Penguin Books in association with New Left Review, 1975.

Maspero, H., *Les Religions Chinoises*, Paris, Musée Guimet, 1950.

Mathijsen, B., 'Measures to Control Production and Trafficking of Opium and Derivatives in the "Golden Triangle" ', paper presented for the Delegation of the European Communities to South-East Asia, 1982.

Matisoff, J., 'Linguistic Diversity and Language Contact', in J. McKinnon and W. Bhrukrasri (eds.), *Highlanders of Thailand*, Kuala Lumpur, Oxford University Press, 1983.

Meesuk, O., 'Income, Consumption and Poverty in Thailand 1962/3 to 1975/6', *World Bank Working Paper No. 364*, Washington DC, 1979.

Merleau-Ponty, M., 'Le Doute de Cezanne', *Fontaine*, 14, December 1947.

Miles, D., 'Some Demographic Implication of Regional Commerce: the Case of North Thailand's Yao Minority', in G. Ho and E. Chapman (eds.), *Studies of Contemporary Thailand*, Canberra, Australian National University Press, 1973.

Moerman, M., 'Ethnic Identification in a Complex Civilization: Who are the Lue?', *American Anthropologist*, 67, 1965.

———, 'Being Lue: Uses and Abuses of Ethnic Identification', in J. Helm (ed.), *Essays on the Problems of the Tribe*, Proceedings of the 1967 Annual Spring Meeting of the American Ethnological Society, University of Washington, 1968a.

———, *Agricultural Change and Peasant Choice in a Thai Village*, Berkeley and Los Angeles, University of California Press, 1968b.

Moody, E. H., *Sam Pollard*, London, Oliphants, 1956.

Moréchand, G., 'Principaux traits du Chamanisme Meo Blanc en Indochine', *Bulletin de l'École Francaise d'Éxtrême Orient*, XLVII, 1955.

_____, 'Le Chamanisme des Hmong', *Bulletin de l'École Francaise d'Éxtrême Orient*, LXIV, 1968.

_____, 'Etymologie de "rever" dans les langues Miao-Yao', *Langues et Techniques: Natur et Science*, Tome II, Paris, 1972.

Morell, D. and Chai-Anan Samudavanija, *Political Conflict in Thailand: Reform, Reaction, Revolution*, Massachusetts, Oelgeschlager, Gunn and Hain, 1981.

Morse, H., *The International Relations of the Chinese Empire*, London, Longman, 1910.

Moseley, G., *The Consolidation of the South China Frontier*, Berkeley, University of California Press, 1973.

Mottin, J., *Elements de Grammaire Hmong Blanc*, Bangkok, The Siam Society, 1978.

_____, *History of the Hmong*, Bangkok, Odeon Books, 1980.

_____, *Contes et Legendes Hmong Blanc*, n.d.(a).

_____, *Allons Faire le Tour du Ciel et de la Terre: Le Chamanisme des Hmong vu dans les textes*, Bangkok, n.d.(b).

_____, *55 Chants d' Amour Hmong Blanc (55 Zaj Kwvtxhiaj Hmoob Dawb)*, n.d.(c).

Naroll, R., 'On Ethnic Unit Classification', *Current Anthropology*, V, 4, 1964.

_____, 'Who the Lue Are', in J. Helm (ed.), *Essays on the Problems of the Tribe*, Proceedings of the 1967 Annual Spring Meeting of the American Ethnological Society, University of Washington, 1968.

Needham, J., *Science and Civilisation in China*, Vol. I, Sections 1–7, Cambridge, Cambridge University Press, 1965.

Needham, R. (ed.), *Right and Left: Essays on Dual Symbolic Interpretation*, Chicago, University of Chicago Press, 1973.

Nisbet, R., *History of the Idea of Progress*, London, Heinemann, 1980.

Odayashi, Taryo, 'Ethnological Remarks on the *T'u Ssu* System in Southwest China', *Minzikugaku Kiniya* (The Japanese Journal of Ethnology), 35, 2, 1970.

Ohnuki-Tierney, E., 'Sakhalin Ainu Time-Reckoning', *MAN*, 1973.

Papet, J-F., 'A propos du Sinxay Legende Nationale Lao, et de son Traitement Oral en Periode Revolutionnarie', *ASEMI*, X, 2–4, 1979.

Parkin, D., 'The Creativity of Abuse', *MAN*, 15, 1, 1980.

Pendleton, R. L., *Thailand, Aspects of Landscape and Life*, New York, Duell, 1962.

Pollard, S., *The Story of the Miao*, London, Henry Hooks, 1919.

_____, *Tight Corners in China*, London, Andrew Crombie, 1921.

Pollard, W., *The Life of Sam Pollard of China*, London, Seeley, Service and Co., 1928.

Potter J., *Thai Peasant Social Structure*, Chicago, University of Chicago Press, 1976.

Ross, P., 'Language and the Mobilization of Ethnic Identity', in H. Giles and B. Saint-Jacques (eds.), *Language and Ethnic Relations*, Oxford and New York, Pergamon Press, 1979.

Roux, H., 'Quelques Minorities Ethnique du Nord-Indochina', *France–Asie*, Saigon, 92–3, 1954.

Ruey Yih-fu, 'Terminological Structure of the Miao Kinship System', *Academica Sinica*, XXIX, 1958.

_____, 'The Magpie Miao of Southern Szechuan', in G. Murdock (ed.), *Social Structure in Southeast Asia*, Chicago, Quadrangle Books, 1960.

_____, 'The Miao: Their Origins and Southwards Migrations', *Proceedings of the International Association of Historians of Asia* (Second Biennial Conference), Taipei, 1962.

Ryan, M., *Marxism and Deconstruction: A Critical Articulation*, Baltimore, Johns Hopkins University Press, 1982.

Saul, J. S., 'The Dialectic of Class and Tribe', *Race and Class*, XX, 1979.

Savina, F. M., 'Consideration sur la Revolte des Miao', *L'Eveil Economique de l'Indochine*, 373, 1924.

———, *Histoire des Miao*, Hong Kong, Societé des Missions Etrangeres de Paris, 1930.

Schafer, E. H., *The Vermilion Bird: T'ang Images of the South*, Berkeley, University of California Press, 1967.

Schanke, D., *Mister Pop*, New York, David MacKay Co., 1970.

Scherman, L., 'Wohnhaustypen in Birma und Assam', *Archiv fur Anthropologie*, N. F. Band XIV, Heft 3, 1915.

Scheuzger, O., *The New Trail: Among the Tribes in North Thailand*, China Inland Mission, Overseas Missionary Fellowship, 1966.

Scott, J. G. and Hardiman, J., *Gazeteer of Upper Burma and the Shan States*, Rangoon, Government House, 1900.

Shiratori, Yoshiro, 'Ethnic Configurations in Southern China', *Minzikugaku Kiniya* (The Japanese Journal of Ethnology), 25, 1966.

Skinner, G. W., '*Akha and Miao* by A. Bernatzik: Review', *Journal of the Siam Society*, 43, Pt. 2, 1956.

Smalley, W., 'The Gospel and the Cultures of Laos', *Readings in Missionary Anthropology*, 3, 3, 1956.

———, 'Cian: Khmu' Culture Hero', *Felicitation Volumes of South-East Asian Studies presented to Prince Dhaninirat*, Vol. I, Bangkok, The Siam Society, 1965.

Smith, H., *The Story of the United Methodist Church*, London, Overseas Missionary Fellowship, 1932.

Southwold, M., *Buddhism in Life: The Anthropological Study of Religion and the Sinhalese Practice of Buddhism*, Manchester, Manchester University Press, 1983.

Steiner, G., *After Babel: Aspects of Language and Translation*, London, Oxford University Press, 1975.

Stern, T., 'Ariya and the Golden Book: A Millenarian Buddhist Sect among the Karen', *Journal of Asian Studies*, 27, 1968.

Stevenson, H., *The Hill Peoples of Burma*, London, Longmans, 1944.

Stuart-Fox, M. (ed.), *Contemporary Laos: Studies in the Politics and Society of the Lao People's Democratic Republic*, New York, St. Martin's Press, 1982.

Suwan, Ruenyote, 'Development and Welfare for the Hill Tribes in Thailand', in P. Hinton (ed.), *Tribesmen and Peasants in North Thailand*, Chiangmai, Tribal Research Centre, 1969.

Tafjel, *The Social Psychology of Minorities*, London, Minority Rights Group, Report No. 38, n.d.

Tan Chee Beng, 'A Legendary History of the Origin of the Yao People', in A. R. Walker (ed.), *Farmers in the Hills: Upland Peoples of North Thailand*, Taipei, Asian Folklore Monographs, 1981.

Tanabe, Shigeharu, 'Ideological Practice within Peasant Rebellions: The Case of Siam around the turn of the Century', in S. Tanabe and A. Turton (eds.), *History and Peasant Consciousness in South East Asia*, Osaka, Senri Ethnological Studies, National Museum of Ethnology, 1984.

Tapp, N. C., 'Thailand Government Policy towards the Hill-dwelling Minority Peoples in the North of Thailand 1959–1976', unpublished MA thesis, University of London, 1979.

———, 'The Relevance of Telephone Directories to a Lineage-based Society: A Consideration of some Messianic Myths among the Hmong from the Ethnographic Literature', *Journal of the Siam Society*, 70, 1, 1982a.

——, 'Notes from an Unfinished Journal', *khaawsaan suun wicaj chaawkhaw* (Tribal Research Centre Quarterly), 6, 3, Chiangmai, 1982b.

——, *The Hmong of Thailand*, Anti-Slavery Society Indigeneous Peoples and Development Series, Report No. 4, London, 1986.

Thai Government, the, *Report on the Socio-Economic Survey of Hill Tribes in Northern Thailand*, Bangkok, Public Welfare Department, mimeo, 1962; published September 1966.

——, *Programme for the Convenience of Thammacarik Monks going as Buddhist Missionaries to the Hill People in Selected Provinces of the North*, Bangkok, 2 February 1968 (Thai).

——, *Report Concerning the Buddhist Mission to the Hill People in the North 1967*, Bangkok, Public Welfare Department, 1968 (Thai).

——, *Statement of the Policy on the Hill People's Development and Welfare*, Bangkok, Public Welfare Department, undated but current in 1979.

Thao, Cheu, *English–Hmong Phrasebook with Useful Wordlist (for Hmong Speakers)*, Washington, DC, Centre for Applied Linguistics, 1981.

——, 'Hmong Migration and Leadership in Laos and in the United States', in B. Downing (ed.), *The Hmong in the West*, University of Minnesota, Center for Urban and Regional Affairs, 1982.

Thaxton, R., 'Modernization and Peasant Resistance in Thailand', in M. Selden (ed.), *Remaking Asia: Essays on the American Uses of Power*, New York, Pantheon Books, 1971.

Thompson, J. R., 'Mountains to Climb', *Far Eastern Economic Review*, 7 March 1968.

——, 'The Mountains are Steppes', *Far Eastern Economic Review*, 11 April 1968.

——, 'Meo Maw: The Burning Mountain', *Far Eastern Economic Review*, 25 April 1968.

Turner, V. Dramas, *Fields and Metaphors: Symbolic Action in Human Society*, Ithaca, Cornell University Press, 1974.

Turton, A., 'North Thai Peasant Society: A Case Study of Rural and Political Structures at the Village Level and their Twentieth-century Transformations', unpublished Ph.D. thesis, University of London, 1976.

——, 'Poverty, Reform and Class Struggle in Rural Thailand', in S. Jones, P. Joshi, and M. Murgis (eds.), *Rural Poverty and Agrarian Reform*, New Delhi, Allied Publishers, 1982.

——, 'Consent, Coercion and Opposition: Problems in the Study of the Social Consciousness of Rural Producers in Thailand', in S. Tanabe and A. Turton (eds.), *History and Peasant Consciousness in South East Asia*, Osaka, Senri Ethnological Studies, National Museum of Ethnology, 1984.

United Nations, *Report of the United Nations Survey Team on the Economic and Social Needs of the Opium-producing Areas in Thailand*, prepared by W. R. Geddes, J. F. O. Phillips, and R. J. Merrill, Bangkok, Government House Printing Office, 1967.

——, 'UN/Thai Programme for Drug Abuse Control in Thailand, UN Division of Narcotics Drugs', *Progress Report No. 6*, December 1975; Geneva, 1976.

——, 'UN/Thai Programme for Drug Abuse Control in Thailand', *Programme Proposal for the Second Phase (1979–83)*, Geneva, 1978.

United States, Hearings before the Sub-Committee on Future Foreign Policy Research and Development, Committee on International Relations, House of Representatives, 94th Congress, 1st Session, 22 and 25 April 1975: *Proposal to Control Opium from the Golden Triangle and Terminate the Shan Opium Trade*, Washington, DC, 1975.

Van der Meer, C. L., 'Rural Development in Northern Thailand: An Interpretation and Analysis', unpublished Ph.D. thesis, University of Croningen, 1981.

Van Gennep, *The Rites of Passage*, London, Routledge Kegan Paul, 1960.

Van Roy, E., *Economic Systems in Northern Thailand*, Ithaca, Cornell University Press, 1971.

Vial, P., *Les Lolos, Histoire, Religion, Moeurs, Langue, Ecriture*, Shanghai, Impr. Mission Catholique, 1898.

Wakeman, F., 'Rebellion and Revolution: The Study of Popular Movements in Chinese History', *Journal of Asian Studies*, XXXVI, 2, 1977.

Walker, A. (ed.), *Farmers in the Hills: Upland Peoples of North Thailand*, Taipei, Asian Folklore and Social Life Monographs, 1981.

Watson, J. L., *Emigration and the Chinese Lineage*, Berkeley, University of California Press, 1975.

Werner, E., *A Dictionary of Chinese Mythology*, Shanghai, 1932; reprinted as *Myths and Legends of China*, Singapore, Graham Brash Ltd., 1984.

Westermeyer, J., *Poppies, Pipes, and People: Opium and Its Use in Laos*, Berkeley and Los Angeles, University of California Press, 1982.

Wiens, H., *China's March into the Tropics: Han Chinese Expansion in Southern China*, Shoe-string Press, 1954; rev. edn., 1970.

Wingate, A. W. S., *A Cavalier in China*, London, Grayson and Grayson, 1940.

Wordsworth, W., *The Lyrical Ballads*, London, 1798.

Worsley, P., *The Trumpet Shall Sound*, London, MacGibbon Kee, 1968.

Yaj, Txoov Tsawb, *Rog Paj Cai*, mimeo., Sayaboury, Y. Bertrais, 1972.

Yangdao, *Les Hmong du Laos Face au Developpement*, Vientiane, Edition Siosavath, 1975.

_____, 'Guerre des Gas: Solution communiste au probleme des minorites du Laos', *Les Tempes Modernes*, janvier 1980.

_____, 'Why did the Hmong leave Laos?', in B. Downing (ed.), *The Hmong in the West*, Minnesota, Center for Urban and Regional Affairs, 1982.

Yangdao and Larteguy, *La Fabuleuse Aventure du Peuple de l'Opium*, Paris, Presse de la Cite, 1979.

Yegar, M., *The Muslims of Burma: A Study of a Minority Group*, Heidelberg, Otto Harrassowitz, 1972.

Yoshiro, S., *Yao Documents*, Tokyo, Kodansha, 1975.

Young, G., *The Hills Tribes of Northern Thailand: A Socio-Ethnological Report*, Monograph No. 1, Bangkok, The Siam Society, 1961.

Yuangrat, W., 'Current Thai Radical Ideology: The Returnees from the Jungle', *Journal of Contemporary South East Asia*, June 1982.

Filmography

Fink, J. and Doua Yang, *Peace has not been made: A Case History of a Hmong Family's Encounter with a Hospital*, colour VT, 25 min., Rhode Island Office of Refugee Resettlement, 600 New London Avenue, Cranston, Rhode Island, USA, 02920, 1983.

_____, *Great Branches*, New Roots: the Hmong Family, colour (16mm film or VT), 42 min., The Hmong Film Project, Commonwealth Avenue, St. Paul, Minnesota, USA, 55108, 1983.

Geddes, W. R., *Miao Year*, ethnographic film: Contemporary Films, McGraw-Hill, 330 West 42nd Street, New York, USA, 10036 (1967/68).

Haley, N., *The Hmong: Traditional Textiles and Music of the Mountain Peoples of*

Laos, film, 16mm, 7 min., 2258 Commonwealth Avenue, St. Paul, Minnesota, USA, 55108.

Lemoine, J. and Moser, B., *The Meo*, film, 35mm, colour, 54 min., Manchester, Granada Television International (1972).

Levine, K. and Waterworth, I., *Becoming American*, film, 16mm, 60 min., New Day Films, Seattle, USA.

O'Neill, P. and Rugoff, R., *The Best Place to Live*, video, 55 min., Providence, Rhode Island, USA (June 1981–August 1982).

Index

Hwang "king": 7576, 78, 94(104 n.11), 95, 98, 123, 124, 125, 126, 129, 131, 18

183